Quoc Hung Pham

Development of a Superconducting Inverter Based on Magnetic Field Triggered Dynamic Resistance

Karlsruher Schriftenreihe zur Supraleitung Band 040

HERAUSGEBER

Prof. Dr. Tabea Arndt
Prof. Dr. rer. nat. Bernhard Holzapfel
Prof. Dr. rer. nat. Sebastian Kempf
Prof. Dr.-Ing. Mathias Noe

Eine Übersicht aller bisher in dieser Schriftenreihe erschienenen
Bände finden Sie am Ende des Buches.

Development of a Superconducting Inverter Based on Magnetic Field Triggered Dynamic Resistance

by
Quoc Hung Pham

KIT Scientific Publishing

Karlsruher Institut für Technologie
Institut für Technische Physik

Development of a Superconducting Inverter Based
on Magnetic Field Triggered Dynamic Resistance

Zur Erlangung des akademischen Grades eines Doktors der Ingenieurwissenschaften (Dr.-Ing.) von der KIT-Fakultät für Elektrotechnik und Informationstechnik des Karlsruher Instituts für Technologie (KIT) genehmigte Dissertation

von Quoc Hung Pham, M.Sc.

Tag der mündlichen Prüfung: 29. Juli 2025
Erster Gutachter: Prof. Dr.-Ing. Mathias Noe
Zweiter Gutachter: Prof. Antonio Morandi, PhD

Bildnachweis Titelseite: ©KIT, Urheber: Markus Breig

Impressum

KIT Scientific Publishing

Karlsruher Institut für Technologie (KIT)
KIT Scientific Publishing
Straße am Forum 2
D-76131 Karlsruhe

KIT Scientific Publishing is a registered trademark
of Karlsruhe Institute of Technology.
Reprint using the book cover is not allowed.

www.bibliothek.kit.edu/ksp.php | E-Mail: info@ksp.kit.edu | Shop: www.ksp.kit.edu

Print on Demand 2025 – Gedruckt auf FSC-zertifiziertem Papier

ISSN 1869-1765
ISBN 978-3-7315-1452-7
DOI 10.5445/KSP/1000183877

Acknowledgements

I would like to express my sincere gratitude to all my colleagues at the Institute of Technical Physics (ITEP) at the Karlsruhe Institute of Technology (KIT) for their invaluable support throughout my time at the institute.

In particular, I am deeply grateful to Prof. Dr.-Ing. Mathias Noe for his continuous guidance, support, and the many insightful meetings and discussions. Without his encouragement and expertise, this work would not have been possible.

My special thanks go to all members of my group, whose contributions created a fantastic working atmosphere, especially during times when I was the only doctoral student in the group. I would also like to thank Andrej Kudymow for his interest in this research and his pragmatic approach to problem-solving. Also, I am grateful to Johann Willms, Dr. Sonja Schlachter, and Rainer Gehring for their assistance in the laboratory and for sharing their valuable knowledge, which proved essential in overcoming many challenges. I also thank Uwe Walschburger for skillfully manufacturing the mechanical parts required for the test setups.

I would like to acknowledge Dr. Alexandra Jung and Karoline Genswein for their help in preparing cross-sections and making cuts in numerous samples and iron cores. Special thanks also to Rainer Nast for his tireless efforts in laser structuring and etching samples, often on very short notice.

Special thanks go to my fellow doctoral students, Annika Uihlein and Mira Wehr, for the enjoyable shared activities, sporting excursions, and, most importantly, or their emotional support and their readiness to listen at any time.

Finally, I am deeply thankful to my parents, Thuy and Hau, my sister Hanh, and my brother-in-law Marcel for their unwavering support over the past decades. Dear Laura, thank you for your understanding and encouragement, especially during the final phase of this work.

Karlsruhe, May 2025

Quoc Hung Pham

Kurzfassung

Diese Arbeit untersucht eine neue Anwendung von Hochtemperatur-Supraleitern (HTS) als Schalter für leistungselektronische Anwendungen, insbesondere durch die Auslegung und den Nachweis eines vollständig supraleitenden H–Brücken–Wechselrichters. Der Schalter selbst nutzt den dynamischen Widerstand, der in HTS-Bändern entsteht, wenn diese sich in einem magnetischen Wechselfeld befinden und gleichzeitig einen Gleichstrom transportieren. Dadurch ist es möglich, dass im eingeschalteten Zustand des Schalters (Magnetfeld aus) ein verlustfreier Stromtransport stattfindet, während im ausgeschalteten Zustand (Magnetfeld ein) ein beliebig schaltbarer Widerstand schnell erzeugt werden kann. Zunächst erfolgt eine Einführung in HTS-Materialien einschließlich deren Grenzen für die supraleitende Phase sowie die aktuell bekannten Annahmen und Modelle des dynamischen Widerstands.

Vorexperimente wurden durchgeführt, um das dynamische Verhalten solcher Schalter unter verschiedenen Magnetfeldamplituden und -frequenzen zu charakterisieren. Die experimentellen Ergebnisse wurden dabei durch analytische und numerische Modelle gestützt. Das Kommutierungsverhalten des Stromes zwischen zwei parallelen supraleitenden Schaltern wurde zudem näher untersucht. Diese Ergebnisse flossen in die Entwicklung des Wechselrichters ein. Es wurde gezeigt, dass Gleichstrom in Wechselstrom erfolgreich umgewandelt wird und dass der Betrieb über einen langen Zeitraum hinweg stabil ist. Darüber hinaus konnte der Leckstrom, welcher über die ausgeschalteten Schalter fließt, reduziert werden.

Ein SPICE-basiertes Modell ermöglichte die Herleitung von Skalierungsgesetzen, wodurch Materialanforderungen und Effizienzzusammenhänge für Anwendungen mit höherer Leistung aufgezeigt wurden. Strategien wie der Einsatz breiterer Supraleiterbänder und alternativer Substrate wurden bewertet, um den benötigten Anteil an supraleitendem Bandmaterial zu verringern. Zusätzlich wurde eine neuartige Methode zur effizienten Erzeugung langer Magnetfelder auf Basis des Rectangular disk-up-down-assembly (rDUDA)-Prinzips entwickelt und experimentell validiert.

Die Ergebnisse stellen einen ersten Machbarkeitsnachweis dar, zeigen die Realisierbarkeit auf und liefern Werkzeuge für die zukünftige Optimierung und Skalierung. Die Technologie befindet sich derzeit auf einem Technologie-Reifegrad (TRL) von 3-4. Weitere Forschung wird sich darauf konzentrieren, den Wechselrichter zu verbessern, Verluste zu minimieren, Fertigungsprozesse weiterzuentwickeln und großskalige Demonstratoren zu entwerfen, um den Weg zu praktischen Anwendungen im Niederspannungs- und Hochstrombereich zu ebnen.

Abstract

This work investigates a new application of high-temperature superconductors (HTS) as switches for power electronics, in particular through the design and demonstration of a fully superconducting H-bridge inverter. The switch itself uses the dynamic resistance that builds up in HTS tapes subjected to an alternating magnetic field while carrying a dc transport current. This enables lossless current transmission in the on-state and a switchable resistance in the off-state of the switch. At the start, an introduction to HTS materials, including their operational constraints and the mechanisms and models of dynamic resistance are given.

Preliminary experiments were conducted to characterize the dynamic behavior of such switches under various magnetic field amplitudes and frequencies, with experimental results supported by analytical and numerical models. The dynamic behavior of two parallel superconducting switches was also examined more closely. These results were incorporated into the design of an inverter. It has been shown that the conversion of dc current to ac current can be successfully achieved and that the conversion is stable over a long period of time. In addition, the leakage current was reduced.

A SPICE-based model allowed the extrapolation of scaling laws, revealing material requirements and efficiency trade-offs for higher power applications. Strategies such as the use of wider superconducting tapes and alternative substrate materials were evaluated to reduce the amount of superconducting tape required. In addition, a novel stack assembly for efficient magnetization of long HTS tapes based on the rectangular disk-up-down assembly (rDUDA) principle was developed and experimentally validated.

The results represent an initial proof of concept, establishing feasibility and providing tools for future optimization and scaling. The technology is currently at Technology Readiness Level (TRL) 3–4. Further research will focus on improving switch design, minimizing losses, refining manufacturing processes, and designing large-scale demonstrators to move towards practical applications in low-voltage, high-current scenarios.

Contents

Appendix

List of Symbols

A	magnetic vector potential
b,k,B_{ch}	model parameters of eq. (2.6)
B_{a}	amplitude of the external magnetic field
B_{abs}	magnetic field norm
B_{airgap}	air gap field
B_{c}	critical magnetic field
B_{ext}	external magnetic field
B_{op}	operation magnetic field
B_{th}	threshold magnetic field
B_{y}	y-component of the magnetic field
$B_{\parallel}$	parallel component of external magnetic field
$B_{\perp}$	perpendicular component of the external magnetic field
B_0	characteristic magnetic field where $I_{\mathrm{c}}(B_0) = 0.5I_{\mathrm{c0}}$
$B1-4$	magnetic field of the switches S1 – S4
c_0,c_1	model parameters of eq. (3.5)
ΔT	excess temperature
E	electric field
$\bar{E}$	average electric field
E_{c}	critical threshold electric field
f	frequency
f_{avg}	factor for non-linear dynamic resistance
G	geometric mean distance
γ	safety factor
$h_{\mathrm{Ag,top}}$	Ag top film thickness
$h_{\mathrm{Ag,bottom}}$	Ag bottom film thickness

$h_{\mathrm{Au,top}}$	AU top film thickness
h_{buffer}	buffer layer thickness
h_{layer}	layer thickness
h_{rebco}	ReBCO layer thickness
h_{subs}	substrate layer thickness
H	magnetic field strength
η_{el}	electric efficiency
i	current load ratio
i_{in}	input current
i_{leak}	leakage current
i_{out}	output current
i_0	amplitude of the applied dc current
i_{1-4}	branch current of the switches S1 – S4
I	current
I_{c}	critical current
$I_{\mathrm{c,min}}$	minimum critical current
I_{c0}	critical current at self field
I_{coil}	coil current
I_{comp}	compensation current
I_{prim}	primary current
I_{t}	dc transport current
$I_{\mathrm{t,layer}}$	dc layer transport current
$I_{\mathrm{t,sc}}$	dc transport current in superconducting layer
j	current density
j_{c}	critical current density
j_{c0}	critical current density at self field
l	length
l_{g}	rDUDA gap length
l_{mag}	active magnetic field length
L	inductance
μ	permeability
μ_0	free space permeability

n	sharpness of the transition according to power law
N_{comp}	number of primary windings
N_{prim}	number of compensation coil windings
N_{pt}	number of parallel superconducting tapes
ω	angular frequency
P_{dyn}	dynamic power loss
P_{in}	input power
P_{mag}	magnetization power loss
$P_{\mathrm{mag,layer}}$	layer magnetization power loss
P_{out}	output power
P_{trans}	current transport power loss
$P_{\mathrm{trans,layer}}$	layer current transport power loss
P_{total}	layer total power loss
$P_{\mathrm{total,layer}}$	layer total power loss
$P_{\mathrm{dyn,nl}}$	non-linear dynamic loss power
$\dot{q}$	heat flux density
Q_{dyn}	dynamic loss
R	electric resistance
R_{Ag}	resistance of the silver cap layer
R_{Cu}	resistance of the surrounding copper
R_{dyn}	dynamic resistance
$R_{\mathrm{dyn,nl}}$	non-linear dynamic resistance
R_{layer}	layer resistance
R_{load}	load resistance
R_{nc}	normal conducting layer resistance
R_{off}	off-state resistance
$R_{\mathrm{off,min}}$	minimum off-state resistance
$R_{\mathrm{off,npt}}$	off-state resistance of one tape
$R_{\mathrm{p,tot}}$	total parallel resistance
R_{sc}	resistance of the superconducting layer
R_{shunt}	shunt resistance
R_{subs}	substrate resistance

R_{tot}	total resistance
$R1-4$	resistance of the switches
$R1-4_{\mathrm{off}}$	off-state resistance of the switches
ρ	resistivity
ρ_{layer}	layer resistivity
S	surface area
S_{layer}	layer surface area
S_{sc}	surface area of superconducting layer
t	time
t_{f}	fall time
t_{on}	test duration
t_{r}	rise time
t_{sw}	switch pulse width
T	temperature
T	current vector potential
T_{c}	critical temperature
T_{op}	operation temperature
T_{sq}	square wave period
θ	angle of the magnetic field to the c-axis
τ	time constant of exponential decay
u_{1-4}	potentials at various positions in the circuit
v	voltage
v_{in}	input voltage
v_{out}	output voltage
V	voltage
$V_{\mathrm{max,ideal}}$	maximum ideal output voltage
$V_{\mathrm{min,ideal}}$	minimum ideal output voltage
V_{out}	output voltage
$V_{\mathrm{out,ideal}}$	ideal output voltage
$V_{\mathrm{out,meas}}$	measured output voltage
w	half width of the superconducting tape
w_{g}	rDUDA gap width

1 Motivation and Scope of Work

The advantages and technical feasibility of superconductors in energy applications, like cables, rotating machines, transformers, and superconducting magnetic energy storage (SMES), were already shown in many studies and demonstrators over the last decades. A possible option to cover the increasing global demand for energy is utilizing superconductors with their higher efficiencies and power densities compared to conventional technologies. One of the most significant advancements in superconducting power transmission is the alternate current (ac) cable "AmpaCity" in Essen, which, at 1 km, was the longest superconducting cable installed in an operational power grid at that time. It has been in operation for over 7 years without any major incidents [Ste16, Her16, Her19].

Due to the unique properties of superconductors, they also enable new fields of application that are not possible with conventional technology, such as fault current limiters. They can limit the current in the event of a short-circuit in the grid, providing more stability and further expansion opportunities.

Another application that uses the highly non-linear behavior of superconductors, but that has not yet been investigated to such an extent is rectifiers and inverters using superconducting switches. Proposals for a superconducting rectifier and the superconducting switch itself have already been investigated in a number of previous projects [tKat81, vdKlu81a, vdKlu81b, Gen16, Gen21, Gaw19b, Gaw18, Leu22, She93, Wan22]. The switches are mostly utilized for operating superconducting magnets in persistent current mode, having a zero resistance short-circuit for the magnetization current. Superconducting rectifiers in the form of a half–wave and full–wave rectifier have already been investigated in flux pumps for charging magnets.

The scope of this work is to design and build a fully superconducting H-bridge power inverter with fast acting high-temperature superconductor (HTS) switches and perform tests on a laboratory demonstrator as a proof of concept. The switches are magnetic field triggered and based on the effect of the dynamic resistance, which enables fast switching. A circuit diagram of such an inverter is shown in fig. 1.1. The circuit consists of an input, output, and four switches S1 to S4. In inverter mode, the input voltage and current are direct current (dc). The switches are then switched on and off in such a pattern that the current is converted to an ac current.

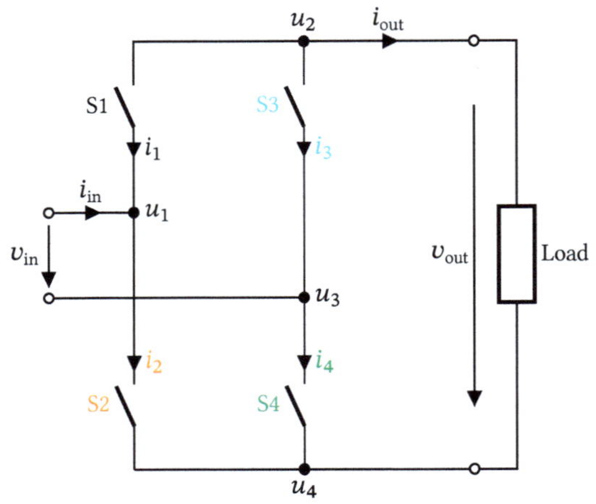

Figure 1.1: Equivalent circuit diagram of a H-bridge power inverter.

The basic current pattern is illustrated in fig. 1.2 where the input current i_{in} is displayed in ■ and the output current i_{out} in ■. The switching pattern according to which the individual switches must be switched to achieve the desired output current is shown above.

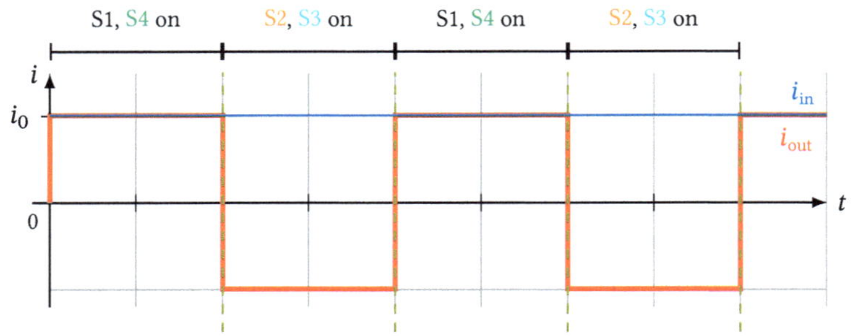

Figure 1.2: Progression of the input and output current of an inverter.

In order to understand the mechanisms of a superconducting switch, the basic properties of high-temperature superconductors are presented in chapter 2. In this chapter, the critical values and dependencies of superconductivity on temperature, magnetic field, and transport current, as well as different methods for increasing the resistance in the superconducting material are introduced.

In chapter 3 a complete characterization of all relevant properties is performed on the superconductor used, and the switch built from it. This includes the magnetic field dependency of the critical current I_c with an external perpendicular field and the temperature-dependent resistance from 77 K to room temperature (RT). The dynamic resistance is measured under various conditions varying the magnetic field amplitude, the frequency, and the dc transport current. First switching experiments are conducted with two parallel superconducting switches and the commutation behavior investigated. The experimental data is then compared with analytical and finite element method models.

In chapter 4 a fully superconducting H-bridge inverter is then designed and afterward a demonstrator is built for a first proof of concept. Various tests with varying input parameters including continuous load are performed to investigate stability and limits of the electrical circuit. Additionally, multiple approaches to improve the switch and circuit are analyzed and applied. Scaling laws are established and up–scaling is carried out taking into account the limits of the superconducting circuit and is investigated by an example use case.

Chapter 5 summarizes all the results and provides an outlook for further work.

2 Fundamentals of High-Temperature Superconductors

Superconductors with their highly non-linear current-voltage behavior are already in use for various power applications. In the superconducting state they can offer zero dc transport current resistance and can carry much higher current densities compared to conventional materials. The distinctive characteristics of superconductors make them an attractive option for improving the efficiency of, or reducing the size and weight of, power devices such as transformers, electric machines, or inverters.

The phenomenon of superconductivity was first observed by Kamerlingh Onnes in 1911, during his investigation of the temperature-dependent electrical resistance of metals with liquid helium (LHe) [Onn91]. He observed that the electrical resistivity of mercury suddenly dropped to a non-measurable value below 4.2 K. In the following years, until 1986, many elements in the periodic table and compounds were found to be superconducting, but only a few were suitable for technical applications [Buc04, Kom95]. These materials require temperatures below 30 K and are therefore also called low-temperature superconductors (LTS). Applications of LTS are limited to research and medical applications due to the costly liquefaction of helium.

This changed with the discovery of a new superconductor class, based on copper-oxides, in 1986 by Bednorz et al. [Bed86]. Soon, a wide variety of materials were found, and superconducting states were possible at temperatures in the range of 90 K, enabling cooling with more economic coolants like liquid nitrogen (LN_2). These superconductors are called HTS. One of the most important superconductors in this class is yttrium barium copper oxide (YBCO), with a critical temperature of 92 K. The rare earth metal yttrium can be replaced by others, forming the subgroup of rare-earth barium copper oxide (ReBCO) superconductors [Kha24, Sha21]. The most important superconductors are displayed in fig. 2.1 with their respective year of discovery and critical temperature.

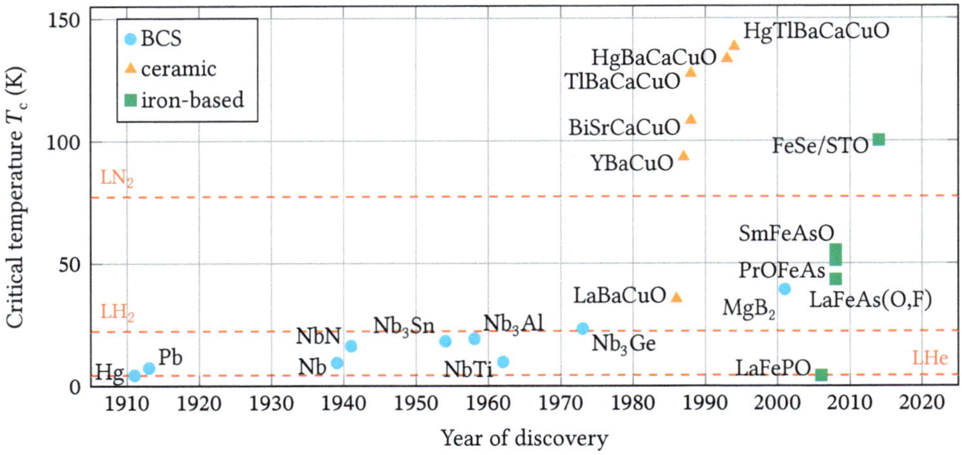

Figure 2.1: Year of discovery for the most important superconductors and their critical temperature T_c. The materials are divided in classic metallic superconductors ●, ceramic copper-oxides ▲, and iron-based superconductors ■.

The theories of superconductivity, like the Bardeen–Cooper–Schrieffer (BCS) theory [Bar57], are already well described in literature [Buc04, Kha24, Wil86]. This is why the following chapter focuses mainly on ReBCO superconductors, which were used in this work. For the application of superconductors in switches, it is crucial to understand their behavior under various external influences, like current, temperature, and magnetic field.

2.1 Critical Parameters of Superconductivity

Superconductors are only in the superconducting phase if certain criteria are not exceeded. These are discussed in the following.

2.1.1 Critical Current Density

Superconductors have a significantly higher current density compared to conventional metallic conductors, but it is constrained by the critical current density j_c or the critical current I_c. Beyond this current, resistance builds up and therefore a voltage drop along the superconductor is measurable. The transition from superconducting to normal conducting state is highly non-linear and is also called the power law which, is stated in eq. (2.1),

$$E = E_c \left(\frac{j}{j_c}\right)^n \tag{2.1}$$

with E as the electric field, i.e. the voltage drop along the superconductor, and E_c the critical threshold electric field. The exponent, also called the n-value, determines how steep the increase in resistance is. For HTS the n-value is within the range of 20 to 40. Since the transition to normal conducting state is continuous, the E_c-criterion has been set to $1\,\mu V\,cm^{-1}$ for HTS materials [IEC06a, IEC20].

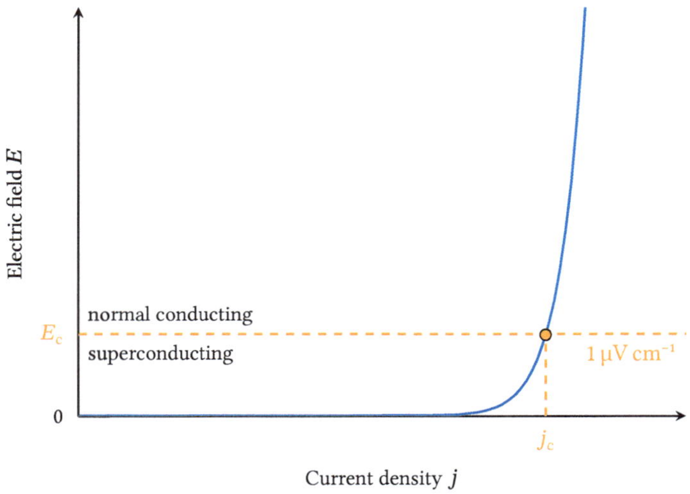

Figure 2.2: Schematic of electric field-current density behavior. The graph shows the power law function with a critical current criterion of $1\,\mu V\,cm^{-1}$ for HTS.

2.1.2 External Magnetic Field

Applying an external magnetic field B_{ext} on a superconductor has a large impact on its properties. The magnetic field decreases the current carrying capacity and the critical temperature T_c. If the magnetic field exceeds the material specific critical magnetic field B_c the superconductor becomes normal conducting. The data in fig. 2.3 show the external magnetic field dependence of the critical current of a typical ReBCO tape. Both the amplitude and the angle of the magnetic field have a major influence on the critical current. The angle θ is defined as the angle between the external field and the perpendicular axis of the superconducting lattice plane. It can be seen that a perpendicular magnetic field $B_{\perp}$ has a more significant influence on the current carrying capacity than a parallel field $B_{\parallel}$. Thus, it is crucial to know the orientation of the superconductor in an external magnetic field, in order to accurately model the behavior in applications.

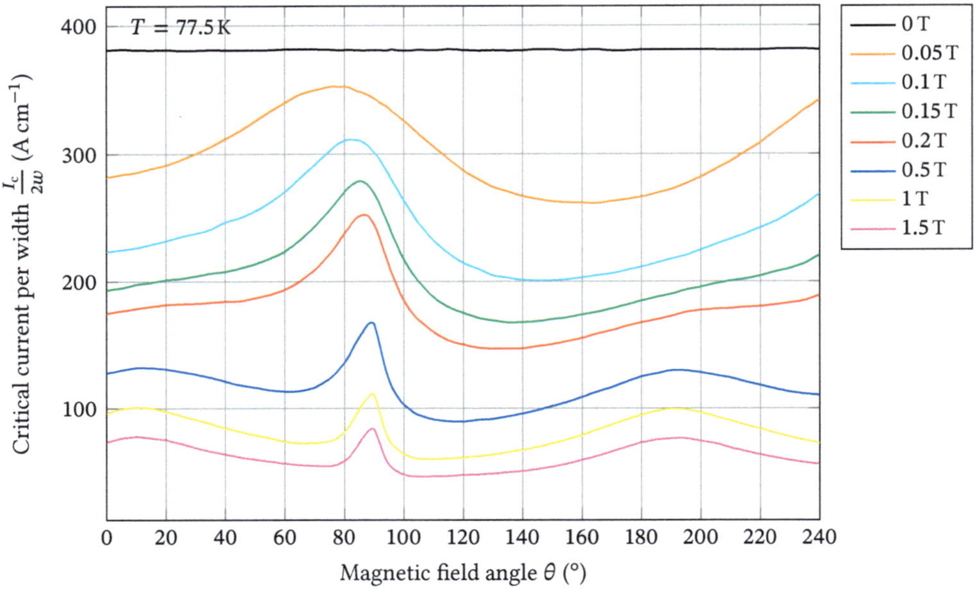

Figure 2.3: Magnetic field angular dependence on the critical current I_c of a ReBCO superconductor. The angle θ is defined as the angle between the external magnetic field and the perpendicular axis (c-axis) of the superconducting plane (see fig. 2.6), (based on data published in [Wim17a], licensed under CC BY 4.0).

2.1.3 Critical Temperature

As already shown in fig. 2.1, a superconducting material has a critical temperature T_c below which the material is in the superconducting state. Figure 2.4 shows the temperature-dependent resistance behavior ($R(T)$-behavior) of a HTS material. Above the critical temperature T_c, the material is normal conducting, with a linear dependence on temperature. Below this temperature the resistance steeply decreases and can be considered zero for dc currents. The critical temperature can be measured by cooling down the superconductor to a low temperature and letting it warm up slowly while continuously measuring the electrical resistance of the superconducting sample. Optimized measurement methods for the $R(T)$-behavior are well described in the literature and standards [IEC06b, Eki06, Ber11]. Above the critical temperature T_c, the resistance behavior and the values is defined by the normal conducting components of the tape and can be calculated according to their cross-section and temperature-dependent resistivity.

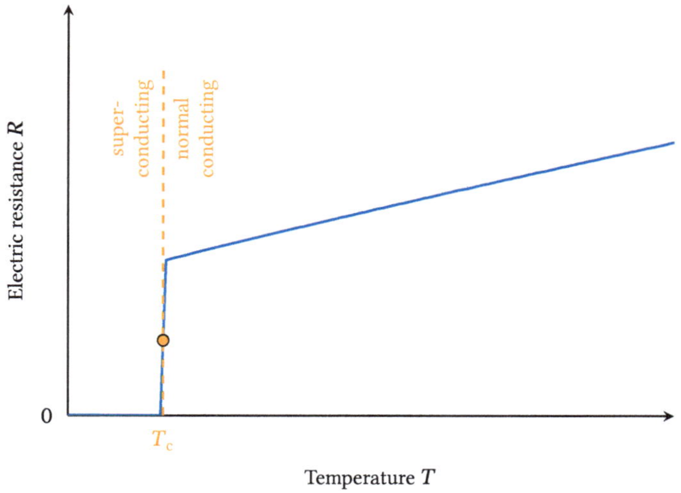

Figure 2.4: Temperature-dependent electric resistance of a superconductor. The critical temperature T_c marks the transition between superconducting and normal conducting states.

Combination of Various External Factors

The critical variables mentioned above cannot be considered in isolation. They are interdependent as illustrated in fig. 2.5. In order to define the operating point in a superconductor use case, all three critical values need to be determined. The shown surface represents the boundary between superconductivity and normal conduction. One exemplary operating point indicated by ● is illustrated. This point represents the maximum current density at the given temperate T_{op} and external magnetic field B_{op}. When designing superconducting power applications, it is important to ensure that this surface is not exceeded under any circumstances, as this would result in a transition to the normal conducting state.

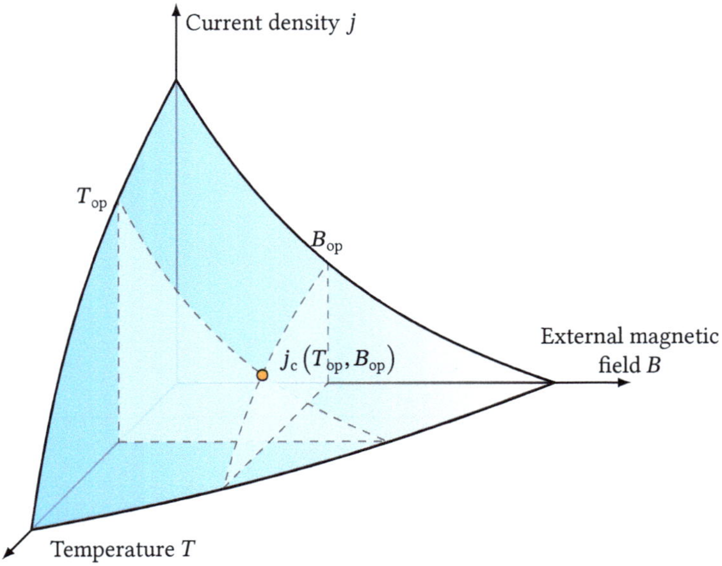

Figure 2.5: Illustration of the limiting factors for superconductivity. The current density j, temperature T, and external magnetic field B_{ext} are interdependent. Every operating point below the surface in ■ is in superconducting state.

2.2 Second Generation High-Temperature Superconductors

The properties of second generation HTS, also called ReBCO, are now described in more detail, since only these are within the scope of this work.

2.2.1 Structure of Rare-Earth Barium Copper Oxide Tapes

The superconducting material ReBCO on its own is very brittle and resembles a ceramic. In order to use these materials in technical applications, sufficient mechanical, electrical, and thermal stabilization is required. The typical structure of a thin-film ReBCO tape is shown in fig. 2.6. Additionally, the whole tape consists of multiple layers of different materials, each with its own purpose.

Furthermore, the base consists of a metallic substrate material (■) such as Hastelloy® or a nickel-tungsten alloy which gives the tape mechanical strength and is typically 20 µm to 100 µm thick.

The superconducting material (■) is separated from the substrate by one or more thin buffer layers (■). These layers are essential to prevent any diffusion or reactions between superconductor and substrate. Additionally, the buffer layers enable the crystal growth of the superconducting layer by presetting the texture with the correct lattice, which is essential for the superconducting phase. The thickness is below 1 µm.

The superconducting layer is deposited by a manufacturer-specific thin-film deposition method, such as pulsed vapor deposition (PVD) [Li02], metal–oxide chemical vapor deposition (MOCVD) [Maj15], or chemical solution deposition (CSD) [Dri12]. The current carrying ReBCO layer is typically 1 µm to 3 µm thick. Also, the deposited superconducting layer can contain small, local defects at which an electric field can build up causing hotspots. To compensate this effect, a 1 µm to 2 µm thin silver (Ag) cap layer (□) is deposited as thermal and electric stabilization.

When used in power engineering applications with high currents, an additional electrical stabilization is added. This usually consists of electroplated or laminated copper (Cu) (□). In the event of a fault, the current flowing in the superconductor can commutate to the surrounding copper and thus prevent the destruction of the superconductor due to heat generation. The thickness of the electrical stabilization is usually between 10 µm and 100 µm. Further information about manufacturing of HTS tapes and different thin-film deposition techniques can be found in literature [Sei15, Sha21, Sel09, Iij00, Goy04].

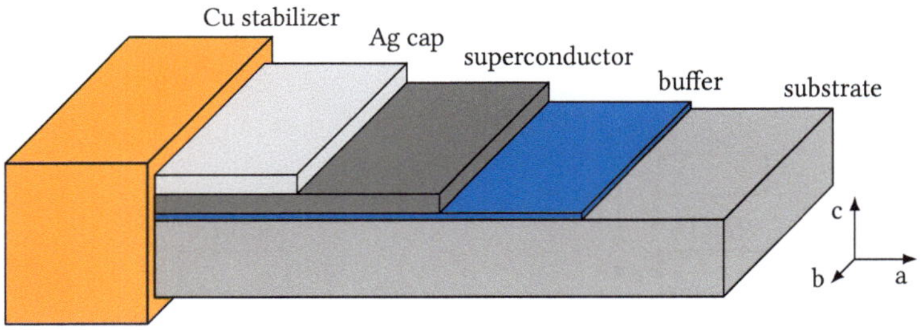

Figure 2.6: Schematic structure of ReBCO tapes.

If the conductor is considered in its normal conducting state, the electrical resistance results from the temperature-dependent specific resistances and the cross-sectional areas of the materials involved in the current transport. In case of a ReBCO tape, the total resistance of the tape R_{tot} can be calculated according to fig. 2.7a as follows

$$\frac{1}{R_{tot}(T)} = \frac{1}{R_{subs}(T) + R_{sc}(T) + R_{Ag}(T) + R_{Cu}(T)} \tag{2.2}$$

with the temperature T and R_{subs}, R_{sc}, R_{Ag}, and R_{Cu} as the temperature-dependent resistances of the substrate, superconducting layer, silver cap layer, and copper stabilization. Due to the high resistance of the buffer layer, the layer does not contribute to the current carrying capacity and can be neglected. The resistance of each layer R_{layer} can be calculated with its thickness h_{layer} and resistivity ρ_{layer}, and the length l and width $2w$ of the superconducting tape.

$$R_{\text{layer}}(T) = \rho_{\text{layer}}(T) \cdot \frac{l}{2w \cdot h_{\text{layer}}} \tag{2.3}$$

All normal conducting layers can be summarized to the temperature-dependent normal conducting resistance R_{nc} as displayed in fig. 2.7b. This simplifies eq. (2.2) to

$$R_{\text{tot}}(T) = \frac{R_{\text{sc}}(T) \cdot R_{\text{nc}}(T)}{R_{\text{sc}}(T) + R_{\text{nc}}(T)} \tag{2.4}$$

with $R_{\text{nc}}(T)$ defined as

$$\frac{1}{R_{\text{nc}}(T)} = \frac{1}{R_{\text{subs}}(T) + R_{\text{Ag}}(T) + R_{\text{Cu}}(T)}. \tag{2.5}$$

When the superconducting layer is in the superconducting phase, the dc resistance of the superconducting layer R_{sc} is zero. However, it is possible to build up a resistance in the superconducting layer with an additional external alternating magnetic field, the so-called dynamic resistance. The resistance of the superconducting layer depends, among other things, on the magnetic field amplitude and frequency $\left(R_{\text{sc}} = R_{\text{dyn}}(B_{\text{a}}, f)\right)$. The dynamic resistance is explained in more detail in section 2.3.

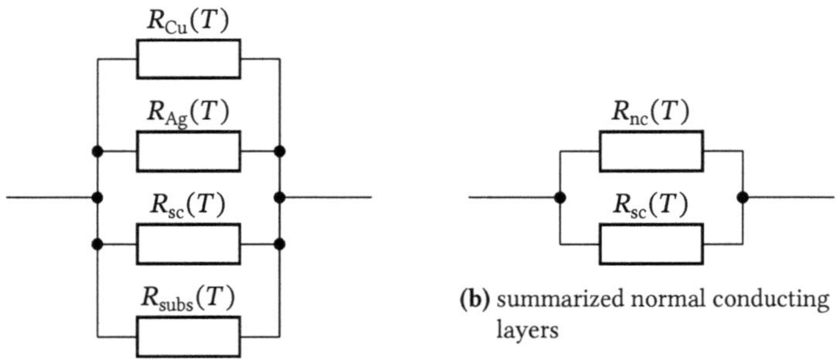

(a) full tape

(b) summarized normal conducting layers

Figure 2.7: Equivalent circuit of total resistance R_{tot} of a superconducting tape.

2.2.2 Critical Current Magnetic Field Dependency

As already stated in section 2.1.2 and fig. 2.3, the critical current I_{c} of HTS strongly depends on the amplitude and orientation of the external magnetic field. These dependencies differ drastically from tape to tape, as manufacturers use different manufacturing processes. A collection of magnetic field-dependent critical currents of different superconductor manufacturers and conductor types can be found in [Wim17b]. In fig. 2.8, the angular magnetic field dependence is plotted for a selection of HTS tapes from the manufacturers SuperPower [Wim17a], Shanghai Superconductor [Wim22], Fujikura [Wim21], and Theva [Wim19]. As can be seen the position of the peak critical current differs strongly. For tape manufactured by SuperPower, the maximum current density is shifted by approximately −4° from 90° and for the superconducting tape from Theva the shift is 6°. This can be attributed to the different production methods used by the manufacturers. The knowledge of the exact superconductor orientation in the magnetic field is essential to calculate the critical current in technical applications.

The angular dependency can be modeled with an elliptical equation given in eq. (2.6) with an error of less than 10 % [Gri14b].

$$j_c(B_a,\theta) = \frac{j_{c0}}{\left(1 + \frac{\sqrt{(kB_a\cdot\sin\theta)^2 + (B_a\cdot\cos\theta)^2}}{B_{ch}}\right)^b} \tag{2.6}$$

With B_{ch}, k, and b as model parameters, B_a the magnetic field magnitude, θ the angle between field vector and c-axis (see fig. 2.6), and j_{c0} the critical current density at self-field. Another approach has been proposed by Nibbio et al. [Nib01] where only the magnetic field component parallel to the c-axis is considered

$$j_c(B) = \frac{j_{c0}}{1 + \frac{|B_\perp|}{B_0}}. \tag{2.7}$$

The parameter B_0 is material specific and is obtained from the I_c-B curve where the critical current is half of the zero field critical current I_{c0}.

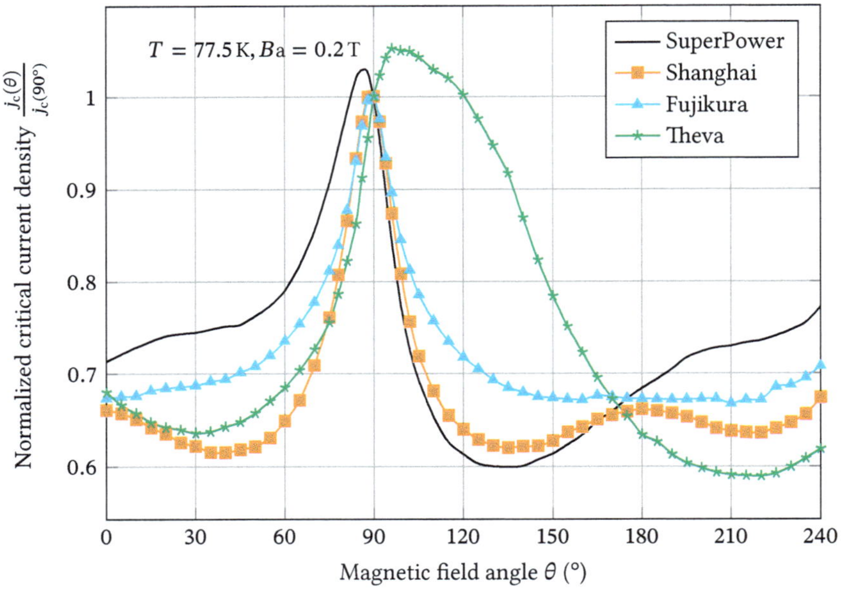

Figure 2.8: Comparison of the angular magnetic field dependency of multiple HTS tapes from different manufacturers at 77.5 K and field amplitude of 0.2 T. The critical current density is normalized to the parallel critical current density $j_c(\theta = 90°)$. Data adapted from [Wim17a, Wim22, Wim21, Wim19], all licensed under CC BY 4.0.

2.2.3 Methods of Resistance Increase

If one or more of the critical values of the superconductor are exceeded, a resistance is built up in the superconducting layer. This current-limiting effect is already utilized by some power applications [dSou23, Li24, Gaw19a] and can be triggered by various causes. A schematic illustration is shown in fig. 2.9.

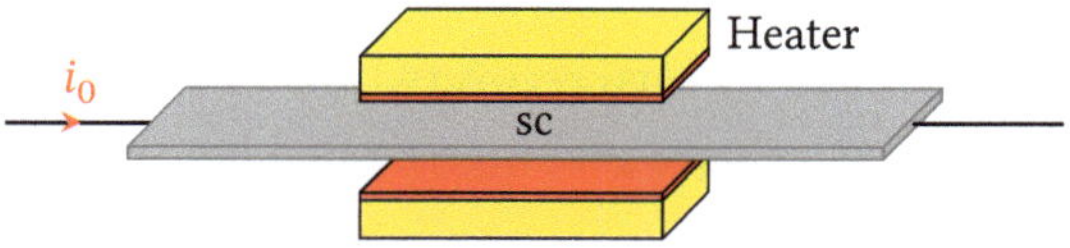

(a) thermally triggered

(b) current triggered

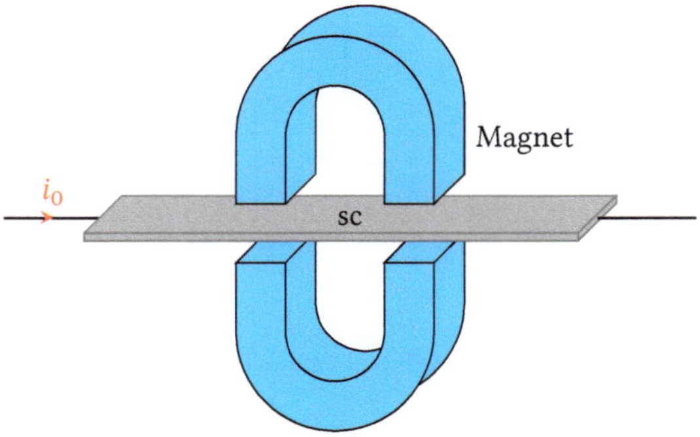

(c) magnetic field triggered

Figure 2.9: Schematic illustration of different methods for increasing the resistance in a ReBCO tape.

2.2.3.1 Thermally Triggered

As illustrated in fig. 2.4, the resistance of an HTS tape increases when the critical temperature T_c is exceeded. The resulting temperature-dependent resistance is defined by the parallel connection of the current carrying layers and can be calculated with eq. (2.2). These kind of thermal switches are utilized in operation of superconducting magnets in persistent current mode like superconducting magnetic energy storage (SMES), magnetic resonance imaging (MRI), and nuclear magnetic resonance (NMR) [Li19, Li24, Tos04]. The persistent switch is connected in parallel to the external power supply for charging the magnet. Then, the superconductor in the switch is heated above T_c, so it becomes resistive. Since the winding of the magnet has no resistance, the current flows only in the magnet, bypassing the switch. When the magnet reaches its operating point, the heater is turned off and the magnet is then short-circuited by the persistent switch putting the magnet in persistent mode. Afterward, the power supply can be turned off and the current flows in a superconducting loop with small or zero joint resistances. The charging state and the persistent mode are shown in fig. 2.10. Since the resistance is defined by the normal conducting layers according to eq. (2.2), a high off-state resistance is possible. A major drawback is the slow switching frequency of 0.1 Hz due to the thermal mass of the materials and their slow thermal recovery time [Oom05, Kim15]. Circuits in form of a half-bridge have also been developed and investigated but these have long response times as well [vdKlu81b, vdKlu81a, Mic17, tKat81, She93].

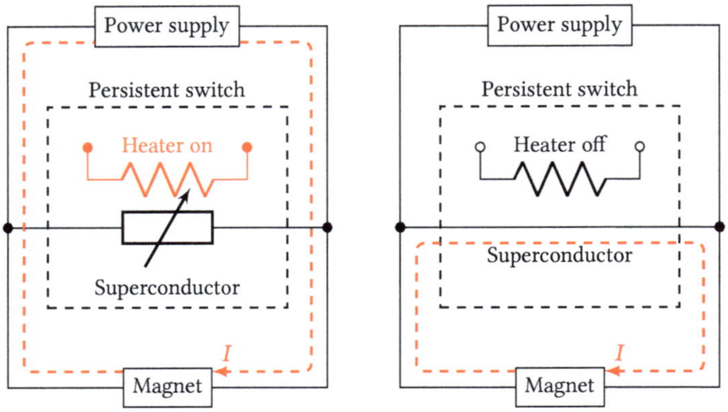

(a) charging **(b)** persistent mode

Figure 2.10: Schematic illustration of different methods for increasing the resistance in a ReBCO tape.

2.2.3.2 Current Triggered

Exceeding the critical current leads to an increase in resistance of the superconductor according to the power law, as shown in fig. 2.2 and eq. (2.1). This effect is applied in resistive fault current limiters (RFCLs) for power grids [Noe07, Kud07, Els12]. A superconducting element is connected in series to the load and is designed in such a way that in every normal scenario the operating current is below the critical current. Therefore, the RFCL has no effect on the impedance of the grid. In the event of a short-circuit, the fault current rises to a value which exceeds the critical current of the superconductor. Consequently, the resistance of the superconducting element rises and increases the impedance of the grid, limiting the fault current. This behavior are shown in fig. 2.11, where the current and the resistance of a RFCL is shown schematically in a fault event. After the short-circuit has been eliminated, the superconductor can recover and the current limiter can be used again. Major design aspects of a RFCL are the maximum allowable fault current and the fault duration in which the superconductor must withstand the fault current. [Sch09, Sch17, Noe23, dSou21]

As the transport current is used as the switching signal, it is not possible to address the switch without changing the transport current. Often, this is not desired and is therefore not suitable as a switch technology for an inverter.

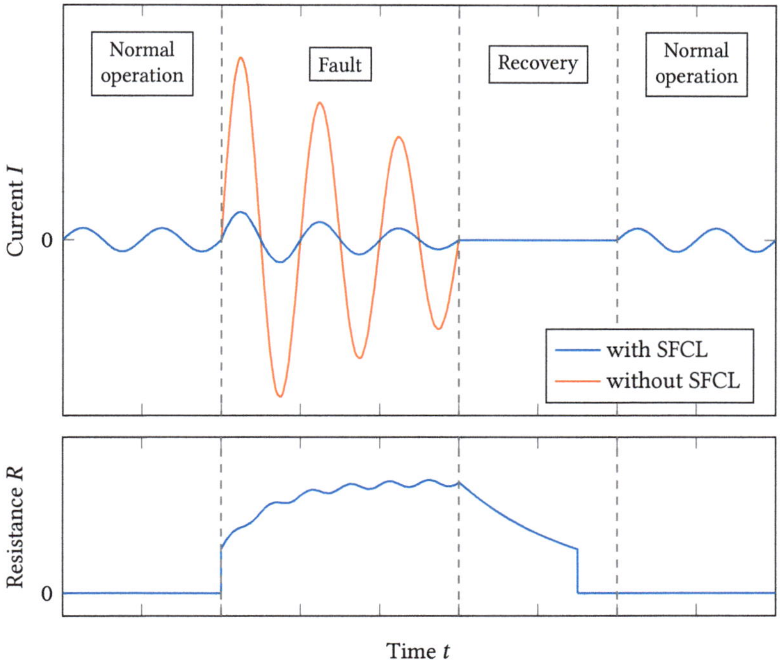

Figure 2.11: Typical limiting process of a RFCL.

2.2.3.3 Magnetic Field Triggered

Another option to force the superconductor into normal conducting phase is to apply an external magnetic field B_{ext}. However, this is difficult to achieve with a constant field. For example, the upper critical magnetic field B_{c2} of YBCO is at approximately 140 T [Smi94]. If an alternating magnetic field is applied, a time-varying resistance builds up along the superconductor. A dc resistance is then obtained by taking the time average. This resistance is also called dynamic resistance R_{dyn} [Oom99, Zha22]. Figure 2.12 shows the magnetic field dependency perpendicular to the ab-plane on the critical current density of a ReBCO tape for dc (●) and ac (■, ▲, ★). The superconducting tape is manufactured by SuperPower with the model number SF12100. For dc field the critical current is decreasing steadily to 20 % of the initial value at 1.1 T. In case of an ac field, the critical current decreases rapidly below 1 % of the initial value at magnetic field amplitudes beneath 0.2 T. The critical current decreases faster with higher frequency of the magnetic field. This experiment shows that an alternating field can be used to significantly reduce the critical current, or a resistance built up. The different influences on the dynamic resistance are discussed in the next section.

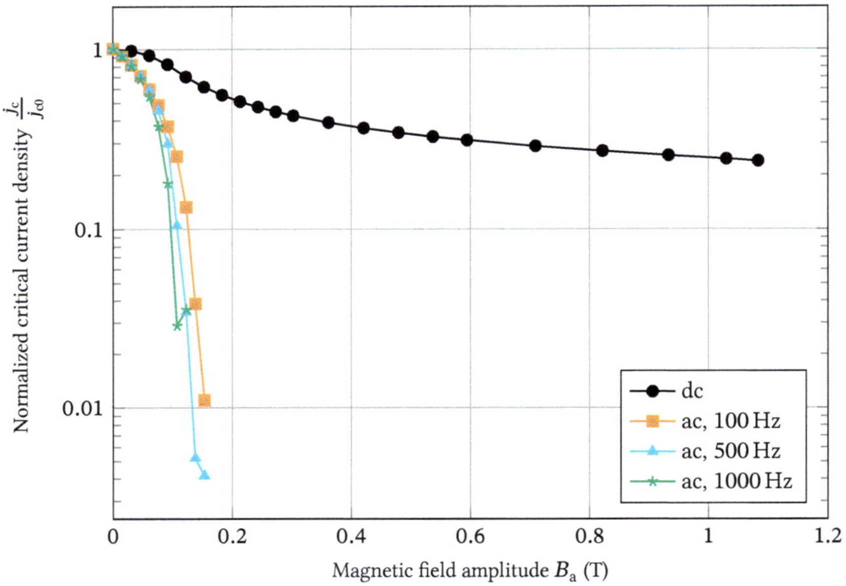

Figure 2.12: Measured critical current of a HTS tape dependent on a perpendicular magnetic field. The dependence on both a dc field and ac fields with different frequencies is shown.

2.3 Dynamic Resistance

Dynamic resistance occurs when a superconductor is subjected to an alternating magnetic field while carrying a dc transport current. This behavior causes losses in superconducting applications where superconductors are subjected to external fields like electric machines or NMR. However, the built-up resistance can also be utilized in flux pumps or in this case superconducting switches for energy applications.

The dynamic losses can be attributed to the flux movements in a superconductor. Applying an alternating magnetic field to a superconductor induces screening currents in the outer region of the superconductor. These shield the inner region of the conductor from the magnetic field creating a flux free region. The dc transport current can flow loss-free in the field-free area. Zero dc resistance is observed. However, when the external alternating magnetic field with a change of rate $\frac{dB}{dt}$ increases over a certain threshold B_{th} and therefore penetrates the inner parts of the superconductor, a voltage is induced across the superconductor according to Faraday's law and a dynamic resistance can be observed [Oom99, Jia17b, Zha22].

2.3.1 Threshold Field Amplitude

Since the external magnetic field B_a needs to penetrate the current carrying area, there is a limit below which the dynamic resistance remains zero. This so-called threshold field amplitude B_{th} for superconducting tapes can be derived from the maxima of the normalized hysteretic ac magnetization loss from Brandt et al. [Bra93]. According to Jiang et al. [Jia17b], this results in

$$B_{th} = 4.9284\frac{\mu_0 j_{c0} h_{rebco}}{\pi}(1-i). \tag{2.8}$$

With the self field critical current density j_{c0} calculated by

$$\frac{I_{c0}}{2wh_{rebco}}, \tag{2.9}$$

the thickness of the superconducting layer h_{rebco} and the dc transport current I_t normalized by the critical current I_{c0}

$$i = \frac{I_t}{I_{c0}}. \tag{2.10}$$

Another approach has been proposed by Mikitik et al. [Mik01] with a non-linear equation for the threshold field amplitude

$$B_{th} = \frac{\mu_0 h_{rebco} j_{c0}}{2\pi}\left[\frac{1}{i}\ln\left(\frac{1+i}{1-i}\right) + \ln\left(\frac{1-i^2}{4i^2}\right)\right]. \tag{2.11}$$

Equations (2.8) and (2.11) have been compared in [Jia17b] with experimental data. At higher current load factors $i > 0.1$ both equations give similar results and eq. (2.8) is sufficiently accurate. At lower current load ratios, only eq. (2.11) can effectively describe the non-linear dependency of the transport current I_t on the dynamic resistance R_{dyn} [Zha19a]. Figure 2.13 shows a comparison of the threshold amplitude B_{th} calculated by eqs. (2.8) and (2.11) showing an approximate agreement at higher load ratios and the high non-linearity and a deviation by a factor of up to 2.5 at smaller current load ratios i below 0.1.

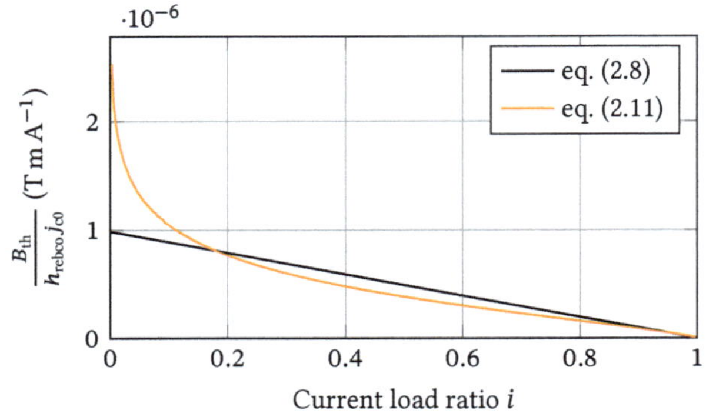

Figure 2.13: Comparison of different equations for the threshold magnetic field amplitude.

2.3.2 Linear Analytic Model

As described in the previous chapter, the dynamic resistance is not only dependent on the transport current. It also depends on the amplitude B_a of the external magnetic field

$$B_{ext}(t) = B_a \cdot \sin(\omega t) \tag{2.12}$$

and its frequency f. The analytic equation has been derived by multiple previous works [Oom99, Mik01, Jia17b] from the critical state model ($n \to \infty$) [Bea64] in which a constant current density j_c for the whole superconductor is assumed. The dynamic resistance can be expressed as

$$R_{dyn} = \frac{4wfl}{I_{c0}} \cdot (B_a - B_{th}) \tag{2.13}$$

with the superconductor width $2w$ and the length l. Respectively, the dynamic loss Q_{dyn} and the dissipated power P_{dyn} are given by:

$$Q_{dyn} = \frac{4wlI_t^2}{I_{c0}} \cdot (B_a - B_{th}) \tag{2.14}$$

$$P_{dyn} = \frac{4wflI_t^2}{I_{c0}} \cdot (B_a - B_{th}) \tag{2.15}$$

Equations (2.13) to (2.15) show a linear dependence of the dynamic resistance on the external magnetic field amplitude and frequency. These equations are widely used and verified with experiments and simulations [Bro20, Jia17a, Ain18, Zha19a] to be sufficiently accurate and can be applied at lower current load ratios and small magnetic field amplitudes. With higher magnetic field amplitudes and load ratios, however, this linear dependence underestimates the dynamic resistance and gives lower values than experimentally measured. Detailed descriptions of the linear equation can be found in literature [Oom99, Oom00, p. 45, Zha22].

A superconducting tape with the parameters in table 2.1 is assumed to compare multiple models at different transport currents and magnetic field parameters. The threshold magnetic field B_{th} is determined by eq. (2.8).

Table 2.1: Parameters of the model tape

Symbol	Parameter	Value
w	half width of the tape	2 mm
h_{rebco}	ReBCO film thickness	1 µm
n	n-value	30
I_t	transport current	3 A to 80 A
B_a	amplitude of the ac field	5 mT to 500 mT
f	frequency of the ac field	3 A to 80 A
B_0	characteristic B-field	0.122 T
E_c	electric field criterion	$1 \cdot 10^{-4}\,\mathrm{V\,m^{-1}}$
μ_0	free space permeability	$4\pi \cdot 10^{-7}\,\mathrm{H\,m^{-1}}$

Figure 2.14 plots the dynamic resistance per unit length for the superconducting tape in table 2.1 according to eqs. (2.8) and (2.13). As the equation states, the dynamic resistance increases linearly with the amplitude and the gradient is the same for all transport currents. Doubling the frequency from 500 Hz to 1000 Hz results in double the resistance. The transport current and therefore also the current load, only has an impact on the threshold field amplitude. The higher the load ratio, the lower the threshold field.

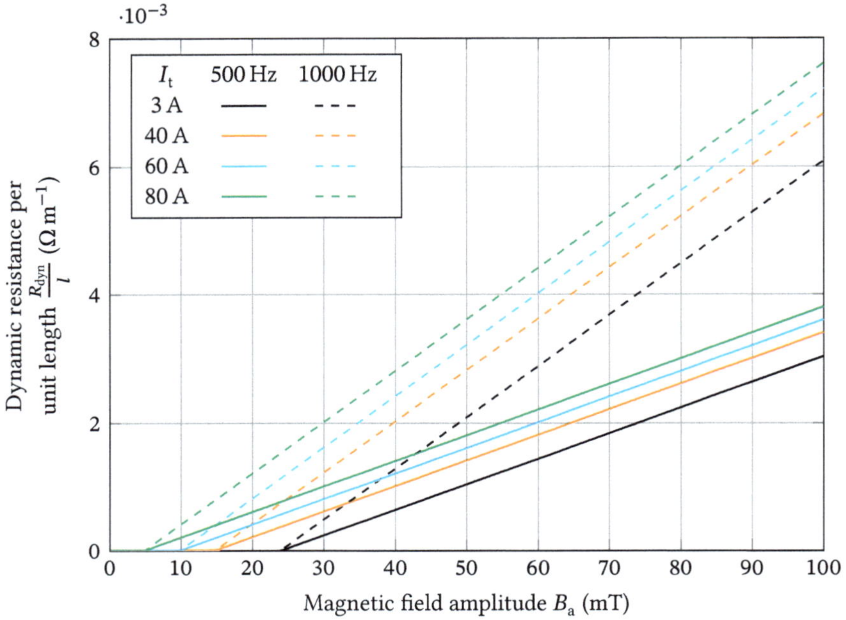

Figure 2.14: Dynamic resistance per unit length according to the linear analytic model from eq. (2.13). A superconducting tape with the parameters from table 2.1 is assumed and the transport current I_t, the magnetic field amplitude B_a, and frequency f are varied.

2.3.3 Non-Linear Analytic Model

In order to be able to describe the non-linear behavior of the dynamic resistance especially at higher magnetic fields and current load ratios, several different ansatzes have been proposed [Oom99, Jia17b, Zha19a]. These incorporate the magnetic field dependency of the critical current (see eq. (2.7)) and the power law with a finite value for the n-value (eq. (2.1)). This is also called time-averaged dc flux flow resistance [Jia17b]. Zhang et al. combined the linear term in eq. (2.13) with the non-linear term to an equation that is valid for low field amplitudes and current ratios as well as higher fields and load ratios near 1 [Zha20a]. The equation can be written as

$$P_{\mathrm{dyn,nl}} = 4wflI_t i \cdot (B_a - B_{\mathrm{th}}) + E_c l I_t i^{n+1} \cdot f_{\mathrm{avg}}(B) \tag{2.16}$$

$$R_{\mathrm{dyn,nl}} = \frac{4wfl}{I_{c0}} \cdot (B_a - B_{\mathrm{th}}) + \frac{E_c l}{I_t} i^{n+1} \cdot f_{\mathrm{avg}}(B) \tag{2.17}$$

where E_c is the critical threshold electric field and n the exponent of the power law in eq. (2.1), and with, when n is even,

$$f_{\text{avg}}(B) = 1 + \sum_{p=0}^{n/2-1} \frac{n!}{(2p+1)!\,[n-(2p+1)]!} \left(\frac{B_\text{a}}{B_0}\right)^{2p+1} \cdot \frac{2^{p+1}\cdot p!}{\pi \prod_{q=0}^{2p+1}(2q+1)} + \sum_{p=0}^{n/2-1} \frac{n!}{(2p+2)!\,[n-(2p+2)]!} \left(\frac{B_\text{a}}{B_0}\right)^{2p+2} \left(\frac{1}{2}\right)^{2p+2} \cdot \frac{(2p+2)!}{[(p+1)!]^2} \tag{2.18}$$

and when n is odd,

$$f_{\text{avg}}(B) = 1 + \sum_{p=0}^{(n-1)/2} \frac{n!}{(2p+1)!\,[n-(2p+1)]!} \left(\frac{B_\text{a}}{B_0}\right)^{2p+1} \cdot \frac{2^{p+1}\cdot p!}{\pi \prod_{q=0}^{2p+1}(2q+1)} + \sum_{p=0}^{(n-1)/2-1} \frac{n!}{(2p+2)!\,[n-(2p+2)]!} \left(\frac{B_\text{a}}{B_0}\right)^{2p+2} \left(\frac{1}{2}\right)^{2p+2} \cdot \frac{(2p+2)!}{[(p+1)!]^2}. \tag{2.19}$$

The linear component covers the area with low current load ratios i and external magnetic fields B_a and the non-linear component covers the area with high i and B_a. A detailed derivation of eqs. (2.16) to (2.19) can be found in [Zha20a, Zha22]. Figure 2.15 displays the dynamic resistance for the superconducting tape from table 2.1 according to eqs. (2.8) and (2.17). Again, the transport current, frequency, and magnetic field amplitude are varied. The y-axis is in logarithmic scale. Since eq. (2.13) includes the linear term from eq. (2.13), both are identical for small current loads, in this case 3 A. However, towards higher transport currents, a non-linear component is added, whereby with higher current load ratio, the magnetic field amplitude, from which the non-linear increase begins, decreases. As in eq. (2.13), the dependence of the dynamic resistance on the frequency is linear.

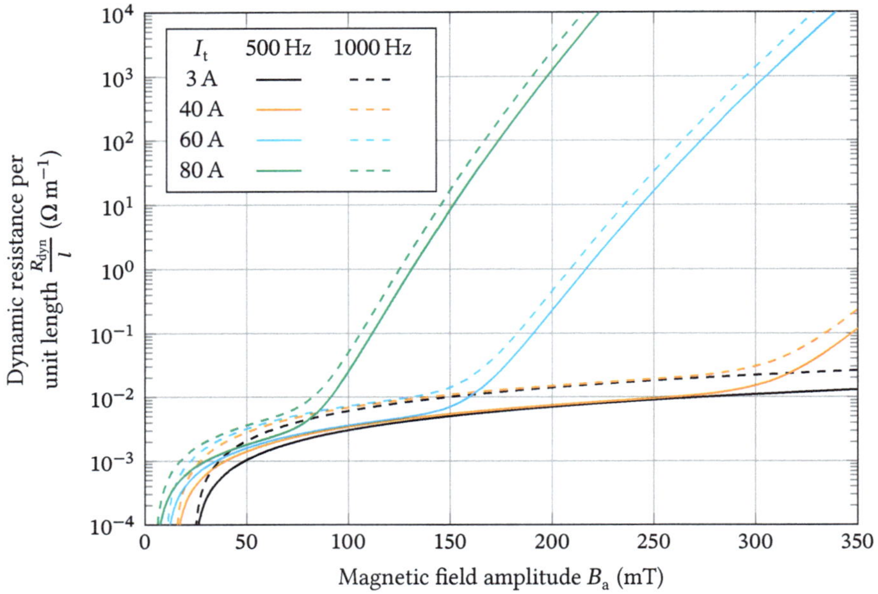

Figure 2.15: Dynamic resistance per unit length according to the non-linear analytic model from eq. (2.17). A superconducting tape with the parameters from table 2.1 is assumed and the transport current I_t, the magnetic field amplitude B_a, and frequency f are varied.

2.3.4 2D FEM Model

In this chapter, the methods for calculating the dynamic resistance are presented. Various methods for calculating Maxwell's equations have been established in the modeling of superconductors in recent years. They differ in how the formulas are resolved and which electromagnetic variables are calculated. Depending on the properties of a domain to be simulated, the resistivity or the conductivity can be used for the calculation in the numerical solver, thus largely effecting the solving time [Gri14a]. The H and the T–A formulations are commonly used for modeling superconductors in two-dimensional (2D) finite element method (FEM). Whereby, the names refer to the electromagnetic potentials used as state variables in the solver.

The H formulation uses the magnetic field H as state variable by combining Faraday's and Ampere's law to

$$\nabla \times (\rho \cdot (\nabla \times H)) = -\mu \frac{\partial H}{\partial t}. \tag{2.20}$$

In the superconducting regions, the resistivity

$$\rho = \frac{E}{j} \tag{2.21}$$

can be derived from power law in eq. (2.1), and for the surrounding domain a large resistivity is assumed, often $1\,\Omega\,\mathrm{m}$. [Gri04, She20, Ain18]

The T–A formulation uses the current vector potential T in the superconducting region and magnetic vector potential A in the whole space. In the superconducting region, the equation

$$\nabla \times (\rho \nabla \times T) = -\frac{\partial B}{\partial t} \tag{2.22}$$

is solved to obtain T which is needed to calculate the current density

$$J = \nabla \times T. \tag{2.23}$$

The magnetic field density B that is needed can be calculated by A with A formulation in the surrounding space and the equation

$$\nabla \times \left(\frac{1}{\mu} \nabla \times A\right) = j. \tag{2.24}$$

The magnetic field can then be calculated from A with

$$B = \nabla \times A. \tag{2.25}$$

In order to simplify the calculation, the $T - A$ formulation utilizes a thin strip approximation which neglects all non-superconducting layers in the superconducting tape and approximates the superconducting layer as a line. This drastically reduces the computation time, however this simplification also removes the information about the influences of the normal conducting layers and the current distribution [Lia17].

Moreover, with the results of the numerical solver and the distribution of electric field E and current density J the ac losses and the dynamic resistance can then be calculated. In general, the losses in a superconducting tape can be distinguished by their power source, the current transport losses P_{trans} and the magnetization losses P_{mag}, in connection with superconductors, these are also called hysteresis losses. The total loss P_{tot} is the summation of all losses of each layer

$$P_{\mathrm{tot,layer}} = P_{\mathrm{trans,layer}} + P_{\mathrm{mag,layer}}. \tag{2.26}$$

In order to calculate each loss component, the total loss is first obtained by integrating the product of current density j and electric field E for each layer

$$P_{\mathrm{tot,layer}} = \iint\limits_{S_{\mathrm{layer}}} E \cdot j \,\mathrm{d}S \tag{2.27}$$

where S_{layer} is the surface area of the respective layer. The transport current loss of each layer is calculated with eq. (2.28) where $\bar{E}$ is the average electric field of the superconducting layer, i.e. the voltage drop per length of the superconducting tape, determined by eq. (2.29).

$$P_{trans,layer} = I_{t,layer} \cdot \bar{E} \tag{2.28}$$

$$\bar{E} = \frac{1}{S_{sc}} \iint\limits_{S_{sc}} E \, dS \tag{2.29}$$

To calculate the voltage drop per length, only the superconducting layer with its surface area S_{sc} is taken into account. The magnetization losses of each layer $P_{mag,layer}$ are now the remaining losses that can be calculated by subtracting the transport losses $P_{trans,layer}$ from the total losses $P_{tot,layer}$

$$P_{mag,layer} = P_{tot,layer} - P_{trans,layer}. \tag{2.30}$$

The relevant dynamic resistance of the superconducting layer R_{dyn} is derived from the transport current $I_{t,sc}$ and the transport loss $P_{trans,sc}$

$$R_{dyn} = \frac{P_{trans,sc}}{I_{t,sc}^2} = \frac{\bar{E}}{I_{t,sc}}. \tag{2.31}$$

2.3.5 Summary

Dynamic resistance in superconductors occurs when a superconductor is in an external alternating magnetic field and simultaneously carries a dc transport current. The level of resistance depends on the material parameters of the superconductor, the gradient of the alternating external magnetic field, and the transport current. However, the dynamic resistance only occurs when a certain magnetic field amplitude, the threshold magnetic field B_{th}, is exceeded and therefore the external field enters the current-carrying area of the superconductor. The threshold field can be calculated with eq. (2.8) or eq. (2.11), the latter being more accurate for lower current ratios.

For calculating the dynamic resistance, multiple analytical equations were proposed to see the dependencies of the variables. The linear analytic model in eq. (2.13) assumes a linear dependency of magnetic field amplitude and frequency on the dynamic resistance. This is reasonably accurate for low magnetic field magnitudes and load current ratios. However, at higher load ratios and magnetic field amplitudes, this equation drastically underestimates the resistance. Another approach which includes the magnetic field dependence of the critical current and the real transition between superconducting state to normal conducting via the power law is denoted in eqs. (2.16) to (2.19). This approach is also accurate at high magnetic fields and load ratios.

With the H and T–A formulation, two of the most frequently used numerical modeling methods and their underlying equations in eqs. (2.20), (2.22) and (2.24) were presented here, whereby the H formulation can be more detailed than the T–A formulation and also allows normal conducting layers to be included in the simulation. Due to the simplifications made, the T–A formulation offers drastically reduced simulation times.

In order to compare the different analytic and numeric models, a superconducting tape with the parameters in table 2.1 is assumed and the dynamic resistance calculated. The threshold magnetic field B_{th} is determined with eq. (2.8). The external magnetic field amplitude B_a and the transport current I_t, therefore also the current load ratio i, are varied and the results of eq. (2.13) (——) and eq. (2.17) (- - -) are displayed in fig. 2.16. The transport currents are 3 A (■), 40 A (■), 60 A (■), and 80 A (■). At lower magnetic field amplitudes both equations agree well with a linear correlation up to a certain point where the non-linear part of eq. (2.17) dominates and the dynamic resistance rises exponentially. The magnetic field from which the non-linear increase begins depends on the transport current I_t or load ratio i. The higher i, the faster it increases exponentially. With a transport current I_t of 3 A, however, no non-linear rise occurs due to the low load ratio i of 0.03 and both equations give the same result.

Additionally, the simulated dynamic resistance based on the T–A formulation from section 2.3.4 is displayed as marks (● , ★ , ■ , and ▲) for the different transport currents. The simulation utilizes the same assumptions listed in table 2.1. The computation time of each set point is approximately 2 min.

In general, the analytical equations are sufficiently accurate for smaller magnetic fields and high load ratios. However, a small deviation can be recognized for small fields. One reason for this may be the use of the linear model in eq. (2.8) for the threshold magnetic field B_{th}, as this underestimates B_{th} at low load ratios i and overestimates it at higher i (see fig. 2.13). The non-linear equation is also valid for higher magnetic fields and can map the exponential increase. At lower current ratios, in this case a transport current I_t of 3 A, the analytical formulas deviate strongly from the simulation results. Especially, at higher magnetic field amplitudes, the analytic equations have a lower gradient.

It should be noted that the models presented in this chapter only calculate the resistance in the superconducting layer. However, for switching operations, the total resistance, introduced with eq. (2.2), is relevant and the other normal conducting layers are to be considered.

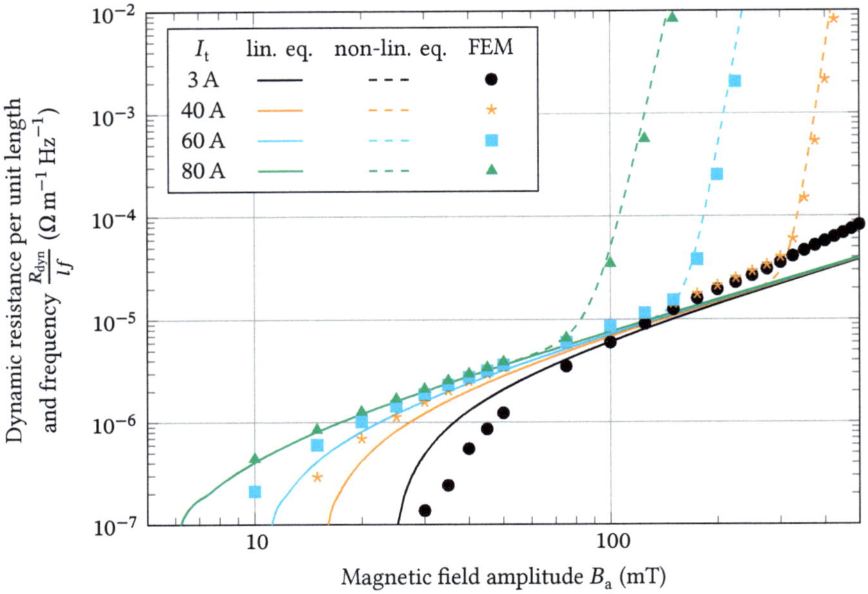

Figure 2.16: Comparison of different dynamic resistance equations. The critical current is constant at 100 A and the transport current I_t and the magnetic field amplitude are varied. The results of the numerical simulation are based on the T–A formulation model of a superconducting tape with the parameters from table 2.1.

3 Experimental Preliminary Investigations

It is important to know the behavior and properties of the superconducting tape in order to design and build a fully superconducting H-bridge inverter. Crucial parameters are the temperature dependent resistance and the magnetic field dependent critical current. In the circuit, the switching time is essentially determined by the resistance that is built up in the switched-off state.

First, the test setup and the equipment used are presented. The magnetic field used to control the superconducting switches is generated by multiple self-wound iron-core electromagnets. Second, the contact-less dc current measurement at cryogenic temperatures is discussed in order to prevent a possible influence on the current distribution by inserted shunt resistors and contact resistances.

The critical current dependency on the magnetic field and the resistance as a function of temperature are measured by a single superconducting tape. The dynamic resistance and therefore the off-state resistance of the superconducting switch is measured under varying magnetic field amplitudes and frequencies. With two superconducting switches in parallel, full redistribution of transport current from one to the other path is shown as first proof of concept. Furthermore, the influence of the dynamic resistance on the commutation behavior, mainly the commutation time, is investigated.

3.1 Test Setup

The experimental setup displayed in fig. 3.1 enables testing of individual superconductor samples with a length of 25 cm in a LN_2 environment. The figure shows a superconducting tape under test with a width of 12 mm with two separate sets of voltage taps, one inside the current feed-in area and one outside. Moreover, the voltage is picked up by beryllium-copper spring contacts. Also, the distances between the voltage taps are 18 cm for the inner and 23 cm for the outer taps. The test current is coupled into the superconductor via two copper contacts pressing against each other over a length of 5 cm. This contact length is sufficient to couple the current completely into the superconducting layer [Eki06, pp. 290–292, 320–324]. Thus, the setup allows accurate resistance measurements via redundant four-terminal sensing eliminating the lead and contact resistance from the measurement [IEC20]. The iron-core electromagnet is mounted in the middle of the test setup to apply an external perpendicular magnetic field over a length of 2 cm on the superconducting tape. Details about the electromagnet can be found in section 3.1.1.

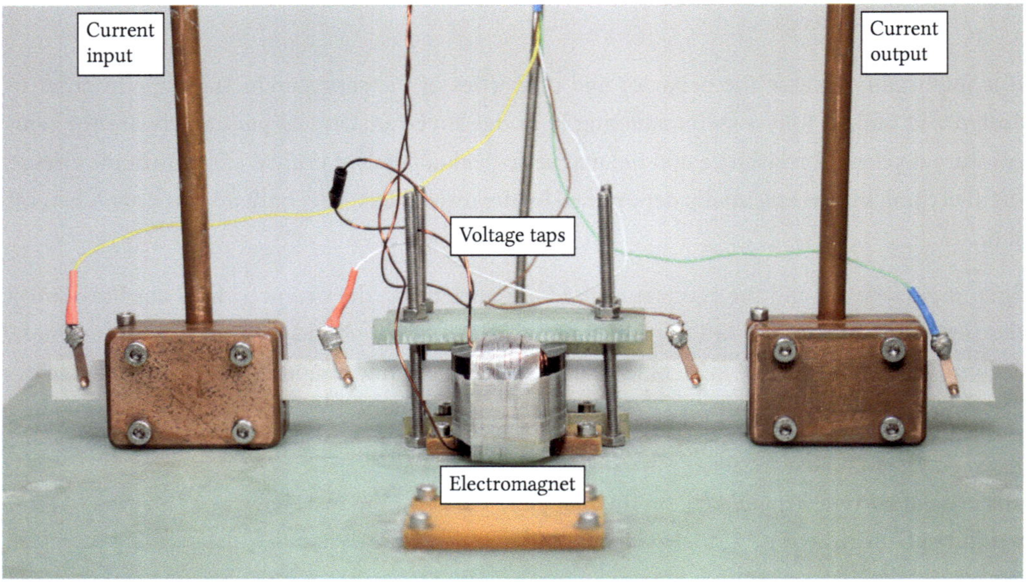

Figure 3.1: Test setup for all single tape measurements. The current is fed in at both ends by two pressure contacts. The voltage is measured inside and outside the current feed-in zone. An iron-core electromagnet is mounted in the middle of the superconducting tape.

The test setup can be adapted to accommodate two parallel superconductors and two independently controllable electromagnets. As shown in fig. 3.2, the setup uses the same current contacts and voltage taps with distances of 23 cm and 33 cm. However, to measure the current of each parallel path without interference, a contactless method is needed. Utilizing shunt resistors would add additional resistance from the shunt itself but contact resistances as well between both superconducting paths. Additionally, both dc, ac, or arbitrary currents in the 10 kHz range, need to be measured. In order to meet these requirements, a closed-loop Hall effect current transducer design is chosen. Two of these current transducers are displayed in fig. 3.2 on the right side. Details can be found in section 3.1.2.

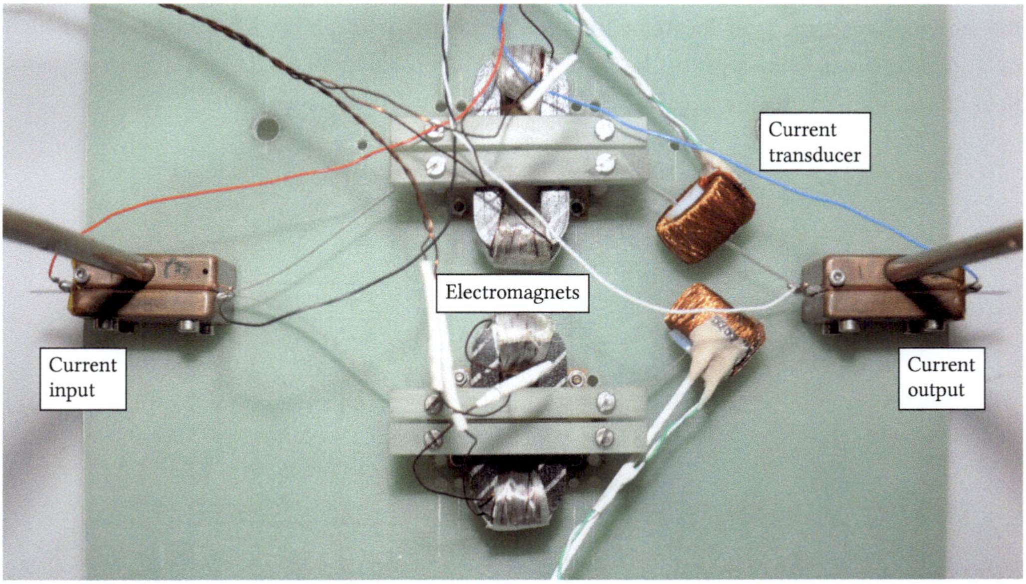

Figure 3.2: Test setup for parallel superconducting tapes. The test setup can be modified to two parallel superconducting tapes, each equipped with a separately controllable electromagnet.

A dc power supply Keysight RP7943A is used to feed current to the superconducting tape in constant current mode. The electromagnets are each individually connected to a bipolar power supply of the type Kepco BOP 20-20 or Kepco BOP 72-6 which are controlled by a four-channel arbitrary waveform generator Tabor WW5064.

For accurately measuring critical currents and other slowly changing voltages, a two-channel nanovolt-meter HP 34420A is used, which is multiplexed by an HP 34970A when more channels are required. Fast changing transient voltages are acquired by an oscilloscope Keysight InfiniiVision DSOX3024A and a 18 bit data acquisition device National Instruments USB-6281 with eight differential input channels. The measurement devices are controlled by software programmed in LabVIEW 2019 and fast transients are recorded with a program written in Python 3.12 with the nidaqmx package [NI25].

3.1.1 Electromagnets

The magnetic field required in this work should be perpendicular to the superconducting tape and as homogeneous as possible. In addition, a high dc magnetic field up to 1 T and ac field with frequencies of up to 1700 Hz and amplitudes in the 300 mT range are to be generated easily. Therefore, the design as shown in fig. 3.3 was chosen. In the test setup, both laminated iron cores are arranged in such a way that an air gap of 1 mm is created between them, in which the superconductor is later located. The high magnetic permeability of the iron allows high magnetic field amplitudes up to the saturation of approximately 1.1 T with a homogeneous field distribution in the air gap.

Parameter	Value
material	E-steel
air gap	2×1 mm
air gap length	2×10 mm
no. of turns	50
saturation field	1.1 T
inductance	1.2 mH

Figure 3.3: Electromagnet for building up a perpendicular magnetic field and triggering the superconducting switches. It consists of two U-shaped laminated iron cores with 50 turns of copper wire in total.

The 50 turns of the excitation coil are chosen because previous calculations show that the iron core saturates at the maximum coil current of 20 A and therefore the operating point of the magnet is in the linear magnetizing section for the entire range. A larger number of turns would only increase the inductance of $\approx$ 1.6 mH which makes it more difficult to produce a fast changing alternating magnetic field due to the higher electromotive force. The coil consists of enameled wire with a diameter of 0.75 mm.

Figure 3.4 shows the characteristic curve of the electromagnet. The current of the excitation current is varied from −20 A to 20 A and the perpendicular magnetic field in the 1 mm wide air gap is measured with a Hall probe of the model HHP-VP # 079 manufactured by AREPOC with a calibrated sensitivity at 77 K of 100.7 mV T^{-1}. The assembly is submerged in LN_2 to create the same conditions as for the later tests and to ensure adequate cooling of the coil. To check for any hysteresis effect, the current is first increased from 0 A to 20 A, then decreased to −20 A and afterward increased to 0 A. Concluding, the graph shows no hysteresis and a symmetric behavior for negative and positive currents with a saturation field of around 1.07 T and the slope between −1 T to 1 T is linear with a gradient of 57.5 mT A^{-1}. The 1 T level is reached with a coil current of 17.4 A.

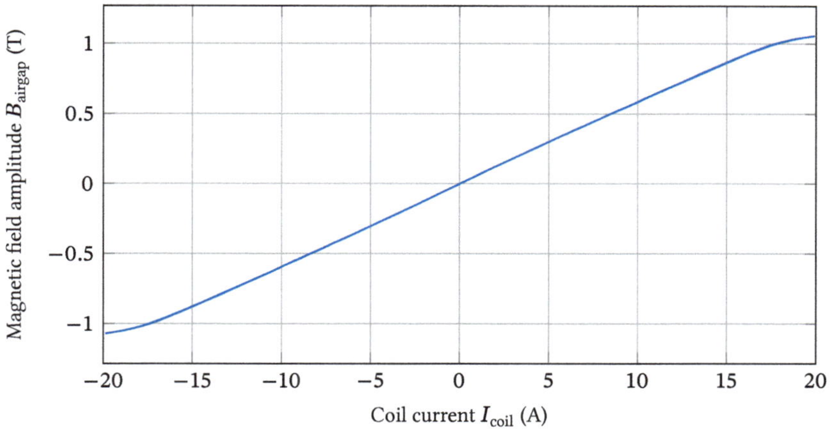

Figure 3.4: Characteristic curve of the electromagnet. Displayed is the air magnetic field B_{airgap} in dependency of the excitation coil current I_{coil}. The measurement is conducted in LN_2 to ensure the same conditions as in later measurements.

3.1.2 Current Transducer

The most common method of measuring currents in superconductor tests is via a shunt resistor. A resistor placed outside the cryogenic area, the value of which is determined very precisely, is connected in series and the voltage drop across it is measured. The voltage is proportional to the current flowing in the superconductor. This method is ideal for single conductors and slowly varying currents. With parallel connected superconductors, however, this method would introduce additional uneven resistances between the superconductors. These have a significant influence on the current distribution of the individual superconductors. A contactless way of measuring current is therefore required, such as closed-loop Hall or fluxgate current sensors.

A schematic of such a sensor is displayed in fig. 3.5. They consist of a ferromagnetic core (■) with a field sensing element (■) inserted into a gap in the core. The core picks up the concentric magnetic field generated by the current flowing through the conductor, i.e. primary winding (■). Afterward, the signal of the magnetic field sensor is processed, and a current is passed on to a compensation coil (■), which generates an opposing magnetic field and cancels the field of the primary winding. Hereby, the compensation current I_{comp} needed for canceling the magnetic field is proportional to the primary current I_{prim} and can be easily measured via a shunt.

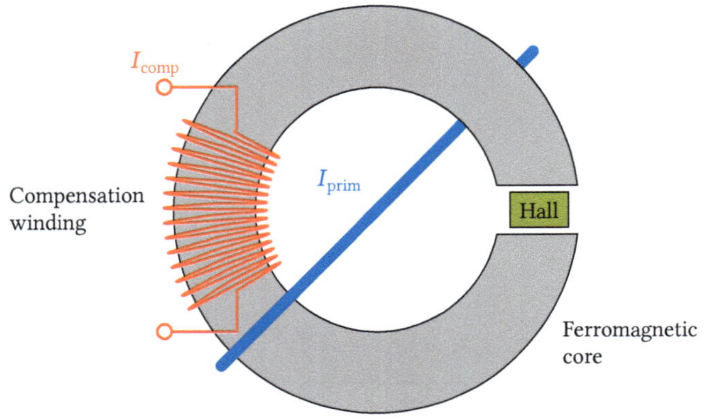

Figure 3.5: Schematic of a closed-loop Hall effect current sensor.

The output voltage V_{out} is given by

$$V_{out} = R_{shunt} \cdot I_{prim} \frac{N_{prim}}{N_{comp}} \tag{3.1}$$

where N_{prim} is the number of turns of the primary coil, in this case 1, and N_{comp} the number of turns of the compensation coil. The advantage of this method is that the magnetic field in the core is compensated to zero and therefore any non-linearities and temperature drifts of the magnetic field sensor are compensated, only the offset remains.

To overcome the offset voltage of the field sensor and the amplifiers, a method called *current spinning* is applied. The excitation current and voltage output terminals of the Hall element are periodically spun in opposite directions. Also, the output voltage is averaged across one period, effectively canceling the Hall and amplifier offset [Mun90, Sch07, Mos17].

Such closed-loop current sensors are widely commercially available, e.g. manufactured by LEM [LEM25] and Vacuumschmelze [VAC25]. However, these all have the electronics for signal processing and provision of the compensation current attached to the core and are therefore not suitable for a cryogenic environment. Specially manufactured current sensors were used for this work, in which the core with Hall sensor and compensation coil are separate from the control electronics, which can be operated at room temperature.

The design input current is ±400 A at an output voltage of ±2.6 V which results in approximately 1500 turns according to eq. (3.1). The resulting current sensor is displayed in 3.6 and consists of a ferromagnetic nano-crystalline VITROPERM® 500 core manufactured by VAC with a size of 30 mm × 25 mm × 15 mm. Furthermore, the 1.5 mm air gap for the Hall sensor and for later threading on the superconducting tape is cut with a computerized numerical controlled cutting wheel. Afterward, the 1500-turn compensation coil is wound by hand with 0.35 mm diameter magnet wire. The Hall effect sensor of the type AREPOC HHP-VP is glued in the air gap. The electronics for signal processing and providing the compensation current of closed-loop current transducers are based on the reference designs of the integrated circuit (IC) DRV411 by Texas Instruments [Tex] with external amplifying stage. Details about the circuit can be found in chapter A.

Figure 3.6: Closed-loop Hall effect current transducer.

In total, six current transducers have been made. Since each compensation coil is wound by hand, deviations in the number of turns are expected and the current transfer function is measured for each coil. This is pictured in fig. 3.7 with the full-scale error for the first two current transducers. Figure 3.7a displays the output voltage for the primary current range from −350 A to 350 A as a linear dependency but with different slopes. The slope, also called sensitivity, is 7.014 mV A^{-1} for the first current sensor and 6.4556 mV A^{-1} for the second. Next, the offset of both transfer functions is below 5 mV. The number of turns can be calculated with the sensitivity and eq. (3.1) to 1426 and 1549. This shows quite a difference compared to the proposed 1500 turns. Table 3.1 lists the sensitivity and the calculated number of turns for all manufactured current sensors.

Figure 3.7b shows the full-scale error for the relevant measurement range and is calculated with

$$\text{full-scale error (\%FSR)} = 100 \cdot \frac{V_{\text{out,meas}} - V_{\text{out,ideal}}}{V_{\text{max,ideal}} - V_{\text{min,ideal}}} \tag{3.2}$$

where the ideal voltage output is taken from the linear fit of each transfer function. The graph shows a similar error curve for both current sensors, with the error being highest at high currents. However, the error remains below 0.027 % over the entire range.

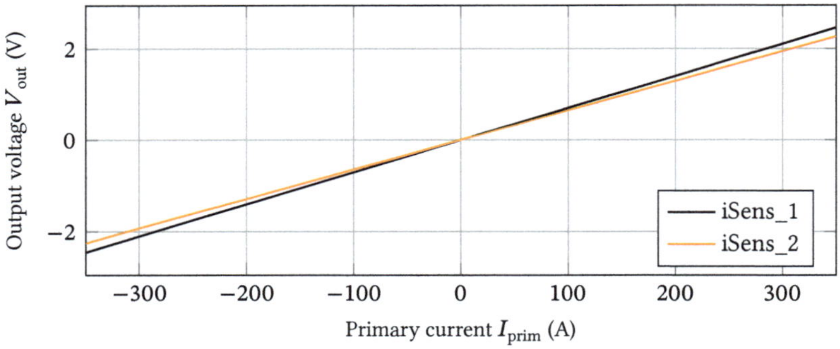

(a) Measured transfer function

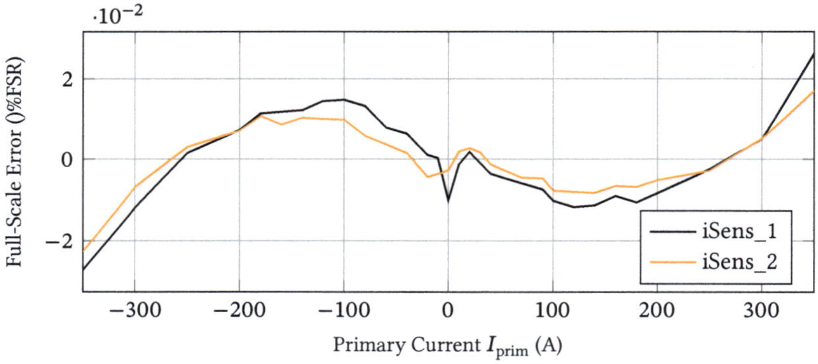

(b) Full-scale error

Figure 3.7: Exemplary dc transfer function of iSens_1 and iSens_2 and full-scale error of the current sensor.

Table 3.1: Sensitivity and calculated number of turns for all manufactured current transducers.

Sensor	Sensitivity (mV A⁻¹)	Calc. no. of turns
iSens_1	7.0140	1426
iSens_2	6.4556	1549
iSens_3	6.5392	1529
iSens_4	6.6267	1509
iSens_5	6.2627	1597
iSens_6	6.7707	1477

3.2 Characterization of Superconductor

Considering sections 2.2.1 and 2.3, achieving high total resistances R_{tot} places multiple demands on the HTS tape in a fully superconducting inverter. Foremost, it is important to have a high normal conductive resistance R_{nc}. A copper stabilization, which is common on commercially available superconducting tapes, would reduce the resistance and is therefore not recommended. The HTS tape should only have a silver cap layer and a substrate that is high-ohmic and as thin as possible in order to achieve high normal conductive resistance R_{nc}. Considering the superconductor specific components from eq. (2.13), the HTS tape should be as wide as possible and have the lowest possible critical current in order to generate a high dynamic resistance R_{dyn}. This is limited by the maximum commercially available width of 12 mm at the present time. However, a high critical current is necessary in order to have enough current capacity for later application.

In this work, the investigated HTS tape is the model SF12100 manufactured by SuperPower Inc. with the advantage of a stabilizer-free design and therefore less normal conducting layers. Only the substrate band and the silver cap layer are contributing to the total resistance R_{tot}. The full specification of the tape can be found in table 3.2. The values are verified by measurements and the thickness of each layer are taken from scanning electron microscope (SEM) imaging, see section B.2. Although thinner substrate thicknesses of 30 µm and 50 µm are commercially available, however, they could not be obtained on short notice, and the superconductor of the selected type was already at hand.

Table 3.2: Parameters of the utilized superconducting tape Superpower SF12100

Symbol	Parameter	Value
w	half width of the tape	6 mm
h_{rebco}	ReBCO film thickness	1.51 µm
h_{subs}	substrate thickness	100 µm
h_{buffer}	buffer film thickness	0.14 µm
$h_{Ag,top}$	Ag top film thickness	1.89 µm
$h_{Ag,bottom}$	Ag bottom film thickness	0.98 µm
$I_{c,0}$	minimal critical current at 77 K, s.f.	338 A
R_{nc} (RT)	normal conducting resistance at RT	3.175 mΩ cm^{-1}
R_{nc} (77 K)	normal conducting resistance at 77 K^{a}	0.729 mΩ cm^{-1}
B_0	characteristic B-field	0.122 T

^{a} obtained by destroying the superconducting phase of the superconducting sample with heat and measuring the temperature dependent resistance.

3.2.1 Magnetic Field Dependent Critical Current

As already described in section 2.1.2, the critical current of a HTS tape is highly dependent on the external applied magnetic field where dc or ac differ greatly. Since in this work a perpendicular magnetic field is applied to the superconducting tape, only this will be considered. The magnetic field dependency can be measured with the test setup in fig. 3.1, where the tape is subjected to a perpendicular magnetic field in a section of 2 cm. Figure 3.8 displays the normalized critical current as a function of the amplitude of the perpendicular magnetic field for dc and ac. With a dc field (●), the critical current decreases steeply, but flattens out with a higher field. At a maximum field of 1.07 T, the critical current is 23.8 % of the original value. In contrast, an alternating field leads to a faster drop in the critical current. For 100 Hz (■), the drop to 25 % is already reached at 107 mT. The higher frequencies 500 Hz (▲) and 1000 Hz (*) result in a faster reduction. These results show that a superconducting switch based on a dc magnetic field is difficult to realize. This is because magnetic field amplitudes which are not trivial to generate, must be achieved to bring the superconductor in normal conducting state at 77 K.

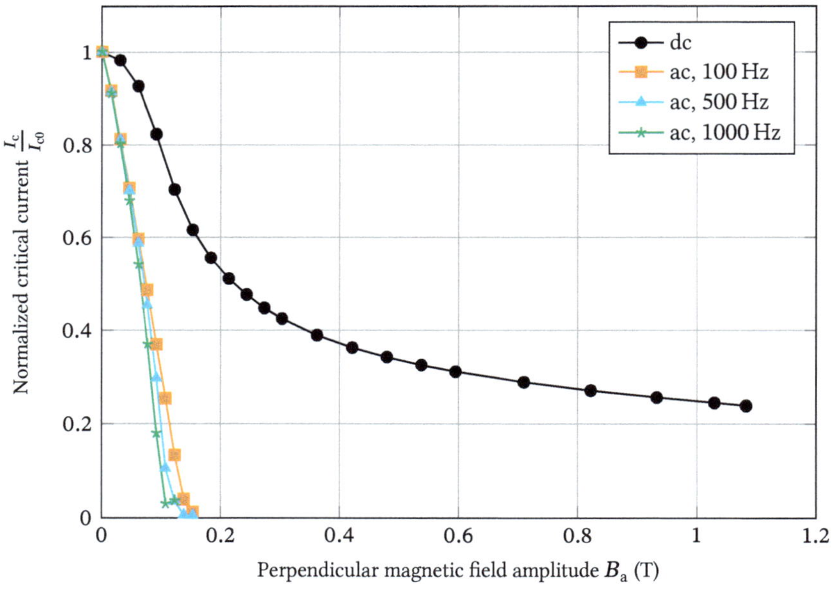

Figure 3.8: Critical current of a HTS tape dependent on a perpendicular magnetic field. The dependence on both a dc field and ac fields with different frequencies is shown.

3.2.2 Temperature Dependent Resistance

The temperature dependent electric resistance is an important performance indicator for designing a superconducting switch. This parameter determines the maximum resistance value of the switch when it is switched off. By measuring the resistance over the temperature, the critical temperature of the superconductor can be determined as already shown in section 2.1.3, but also the resistance of the normal conducting layers according to eqs. (2.2) to (2.5).

The utilized test setup, described in [Ber11, p. 132], consists of a sample holder with a high thermal copper mass and a temperature sensor. This is then cooled with LN_2 down to 77 K and slowly heated electrically back to RT. Meanwhile, the resistance is continuously measured over a distance of 5 cm using four-terminal sensing. The $R(T)$-behavior of the superconductor SF12100 is displayed in fig. 3.9. The sample was then heat treated to destroy the superconductivity of the tape and measured again. Consequently, the superconductor (—) shows a typical behavior, below the critical temperature T_c of 92 K, the resistance is zero. At T_c the resistance rises sharply to $1.05\,\mathrm{m\Omega\,cm^{-1}}$, then rises linearly with the temperature according to the amounts of normal conducting layers. The resistance value agrees well with the theoretical resistance obtained from the individual layer thicknesses in table 3.2 and the resistivities of the normal conducting materials in section B.1, from which the resistance can be calculated using eqs. (2.2) and (2.3). The heat treated sample (—) has a similar curve above the critical temperature T_c, but deviates slightly at higher temperatures. Below T_c, the linear trend continues with a resistance of $0.729\,\mathrm{m\Omega\,cm^{-1}}$ being reached at 77 K which can be split into the silver resistance R_{Ag}/l of $0.785\,\mathrm{m\Omega\,cm^{-1}}$ and substrate resistance R_{subs}/l of $10.25\,\mathrm{m\Omega\,cm^{-1}}$. This corresponds to a resistivity of $9.14 \cdot 10^{-8}\,\Omega\,\mathrm{m}$ in relation to the cross-section superconducting tape as a whole. As the conductivity of silver is orders of magnitude higher than that of the substrate, it determines the total resistance of the superconducting tape when in normal conducting state, although its cross-section as a proportion of the total tape is small.

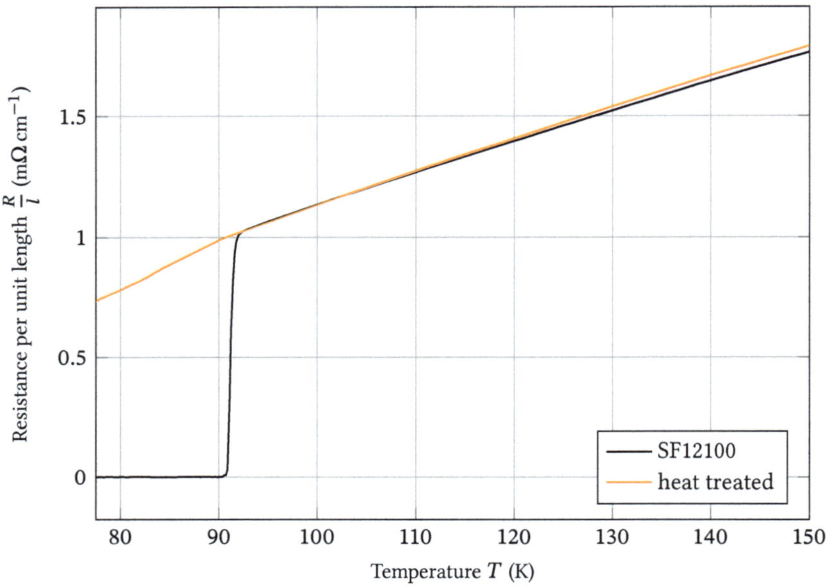

Figure 3.9: Temperature-dependent electric resistance of the superconductor SF12100.

3.2.3 Methods of Increasing the Total Resistance

The temperature-dependent resistance limits for the off-state resistance of the superconducting switch, regardless of how high the dynamic resistance R_{dyn} is. One way to increase the total resistance is to modify the low-ohmic silver cap layer. Figure 3.10 shows different configurations of removal patterns as a schematic representation with the individual layers of the superconducting tape. Configuration A (fig. 3.10a) is the reference sample, the unmodified tape with its resistance in fig. 3.9. In configuration B, the surrounding silver is removed from all four sides in a section of 5 mm, eliminating the parallel resistance of silver from eq. (2.2). The substrate is still connected with the superconductor via the silver surrounding outside the section. In configuration C (fig. 3.10c), the silver layer is removed from the edges and the substrate side on the whole length of the sample. The buffer acts as an insulator between substrate and superconducting layer. This would yield the highest resistance. The theoretical equivalent circuits of the presented configurations are displayed in fig. 3.11. However, it cannot be ruled out that there is no electrical connection between the two layers. It is expected that the total resistance of config. C will be higher than for config. B due to the increased contact resistance.

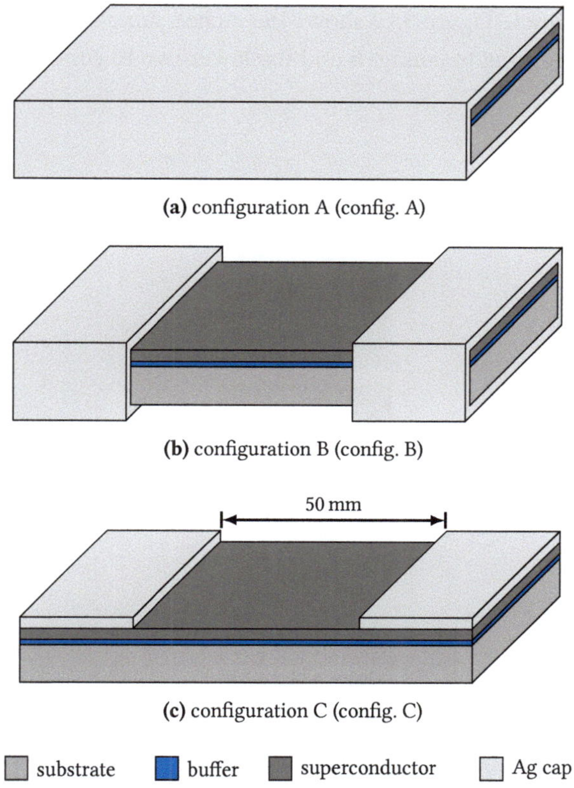

Figure 3.10: Schematic illustration of the etching patterns in order to improve the total resistance.

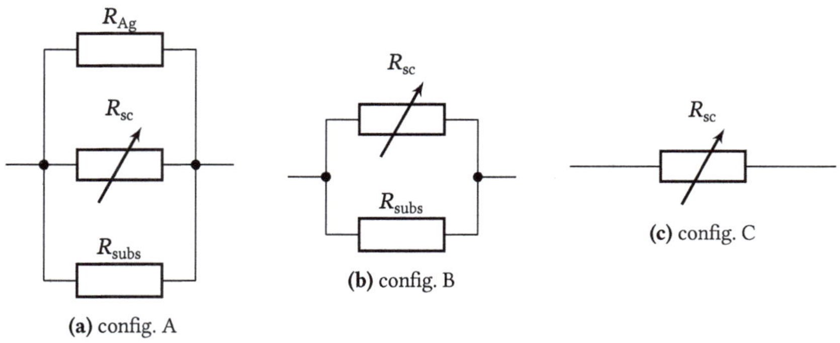

Figure 3.11: Equivalent circuit of the etching patterns in order to increase the total resistance.

The silver layer is chemically removed with a solution of 50 % water, 25 % ammonia, and 25 % hydrogen peroxide by volume. The sections where the silver should not be removed are masked with polyimide tape, and submerged into the solution until the silver is completely removed and the superconducting layer or the substrate layer are visible. The samples are then cleaned in distilled water and ethanol. Figure 3.12 shows the etched samples from top and bottom, the first two are etched according to config. B and the lower two to config. C. The samples are in total 27 cm long.

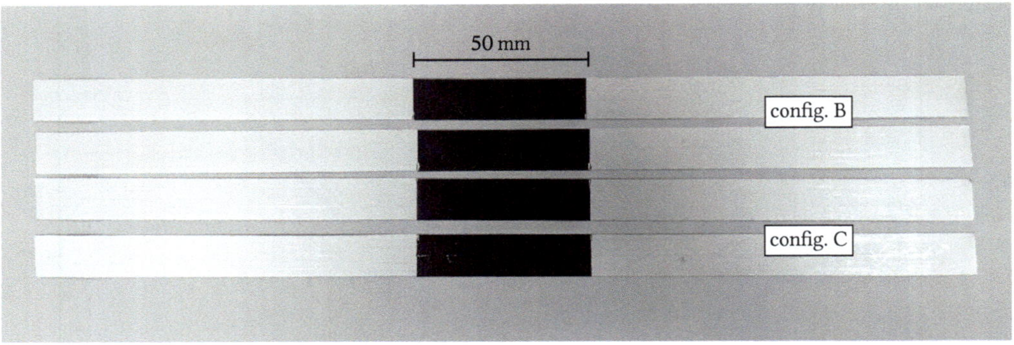

(a) top

(b) bottom

Figure 3.12: Samples after etching from top and bottom side. The top two are etched according to config. B and the bottom two to config. C.

To quantify the increase of resistance of the new modifications, the $R(T)$-behavior is measured again, and the results are displayed in fig. 3.13. The measurement data from config. A are identical to those in fig. 3.9 and are shown there with better scaling. All three curves have a sharp transition at 92 K from superconducting to normal conducting whereby the transition of config. C is stretched over a wider temperature range. The resistance at 110 K is 0.7995 $\mathrm{m\Omega\,cm^{-1}}$ for configuration A, 12.798 $\mathrm{m\Omega\,cm^{-1}}$ for config. B, and 41.182 $\mathrm{m\Omega\,cm^{-1}}$ for config. C, whereby the first two correspond well with the theoretical resistivities and cross-section values. To estimate the resistance at 77 K, a linear extrapolation is carried out with the resistance values of 100 K to 120 K and are displayed as (- - -). The values are 0.7995 $\mathrm{m\Omega\,cm^{-1}}$, 12.154 $\mathrm{m\Omega\,cm^{-1}}$

and 40.060 mΩ cm^{-1} respectively as well. These resistance values are also used as parallel normal conducting resistance R_{nc} to determine the dynamic resistance R_{dyn} from the measured total resistance R_{tot} according to eq. (2.4). In summary, by removing the silver layer the normal conducting resistance can be increased by a factor of 12 for configuration B and 40 for configuration C.

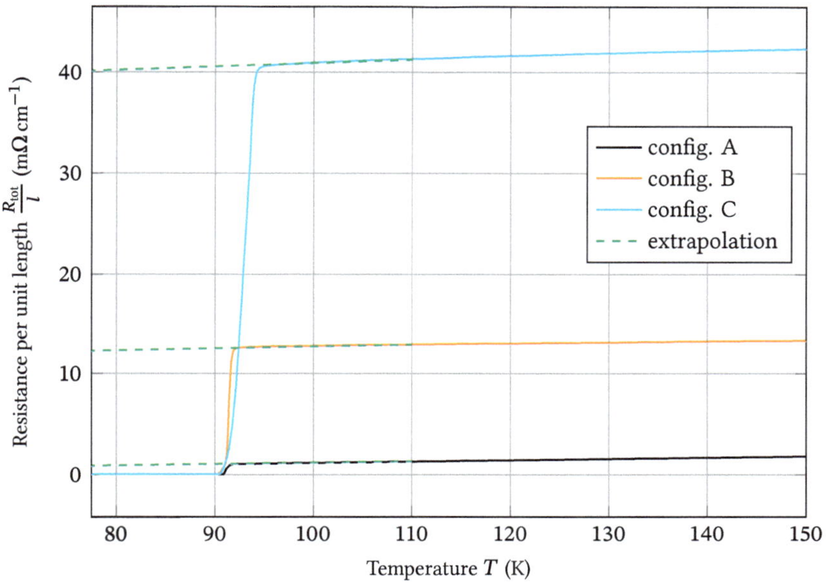

Figure 3.13: Comparison of the temperature-dependent resistance of the different etching patterns presented in fig. 3.10.

In order to check whether the superconductor has been damaged by the etching process, several critical current measurements were carried out, going up to 20 times the value of the criterion. The results in table 3.3 show no significant degradation. In addition, several thermal cycles were run between room temperature and 77 K without any change in the critical current. In order to protect the exposed superconducting layer from external environmental influences such as oxygen or moisture, the etched areas are laminated with 50 µm thick polyimide film on both sides, the result can be seen in fig. 3.14.

Table 3.3: Critical current before and after the etching

Sample	**Untreated**		**Etched**	
	I_c	n	I_c	n
config. A	338 A	34.8		
config. B	341 A	36.2	339 A	35.9
config. C	340 A	35.5	337 A	35.2

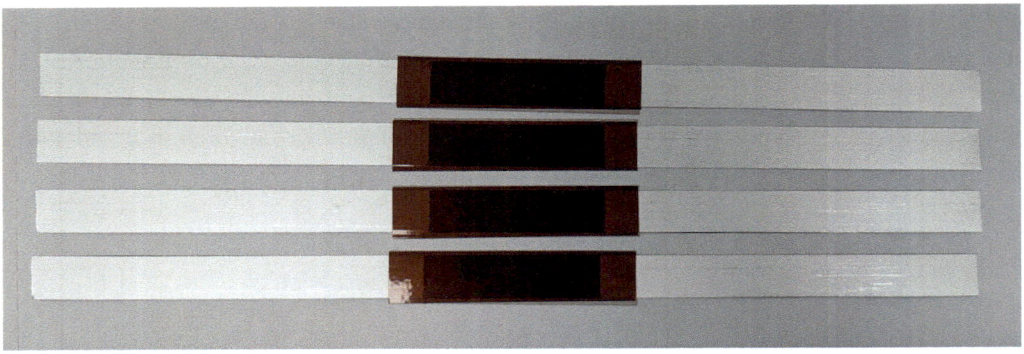

Figure 3.14: Etched tapes laminated with 50 µm polyimide foil on each side.

3.3 Total and Dynamic Resistance

With the fully characterized HTS tape, the dynamic resistance R_{dyn} can now be measured. In particular, the test setup in fig. 3.1 is used, and the power supply is set as a constant current source to the desired dc transport current I_{t}. Afterward, the electromagnet with the predefined frequency and amplitude is toggled on for 5 s during which the voltage drop V_{meas} is measured. The result is averaged over time and is the total resistance R_{tot} of the superconducting tape over the length of the active magnetic field l_{mag}.

$$\frac{R_{\mathrm{tot}}}{l_{\mathrm{mag}}} = \frac{V_{\mathrm{meas}}}{l_{\mathrm{mag}} \cdot I_{\mathrm{t}}} \tag{3.3}$$

3.3.1 Measurement Results

The measured total resistance for config. A without polyimide laminate under varying external magnetic fields from 25 mT to 277 mT is shown in fig. 3.15. The frequency of the magnetic field is set to 500 Hz (top) and 1000 Hz (bottom) and the transport current is 3 A and 20 A. The resistances at both transport currents show a similar linear course, but with a small offset, as expected according to the analytical equations from section 2.3. Also, a threshold field amplitude B_{th} below which no resistance occurs is visible. However, all curves show a drop in resistance above a certain magnetic field; at 500 Hz this is around 250 mT and at 1000 Hz around 150 mT. After this dip, the resistance continues to increase linearly at 1000 Hz. This behavior is likely caused by the transition through different heat transfer mechanisms with pool boiling in LN_2 caused by different heat flux [Mer64, Hel14] and is further investigated in the next section.

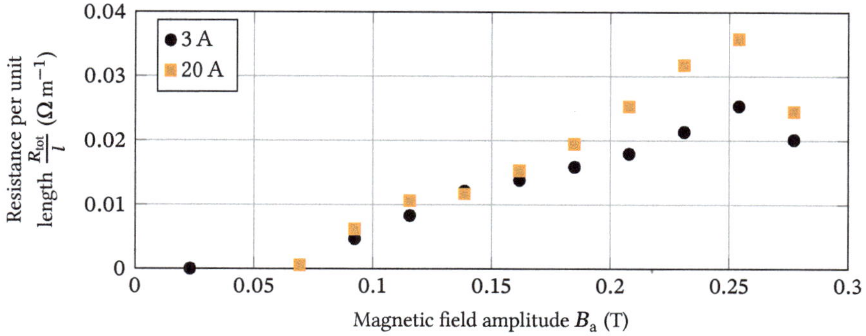

(a) 500 Hz

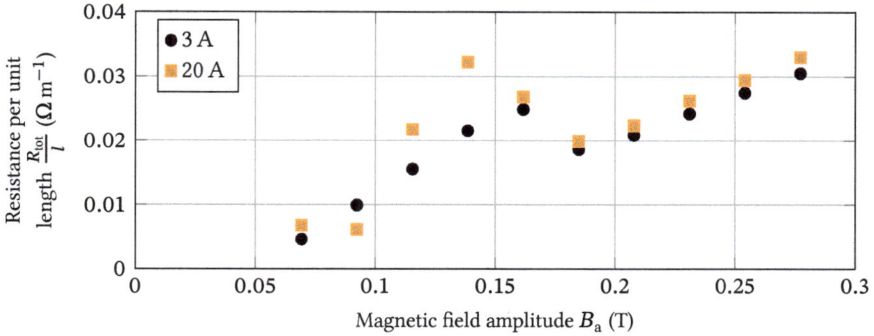

(b) 1000 Hz

Figure 3.15: Measurement of the total resistance R_{tot} of config. A without polyimide laminate.

The dynamic resistance R_{dyn}, as displayed in fig. 3.16, is then calculated with eq. (2.4) and the normal conducting resistances at 77 K from fig. 3.13. The curves show the same curve as the total resistance, with higher magnetic fields a slightly non-linear curve can also be seen. Additionally, the linear analytic eq. (2.13) with the threshold field amplitude from eq. (2.11) is displayed showing a smaller slope than the measurement. Comparing both frequencies, the increase in dynamic resistance is not linear, especially at higher magnetic fields.

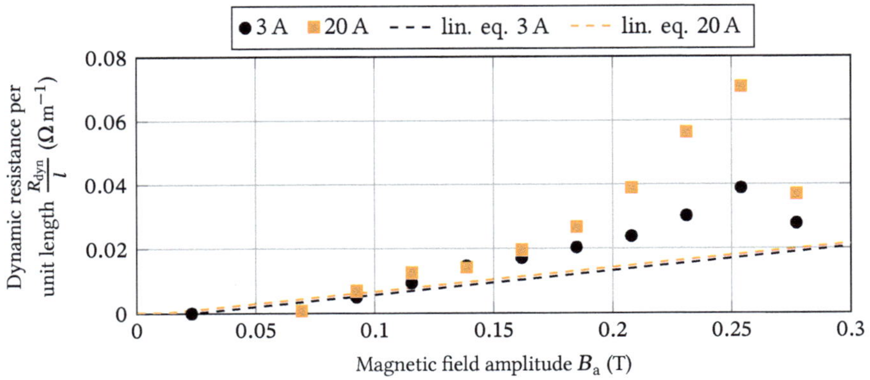

(a) 500 Hz

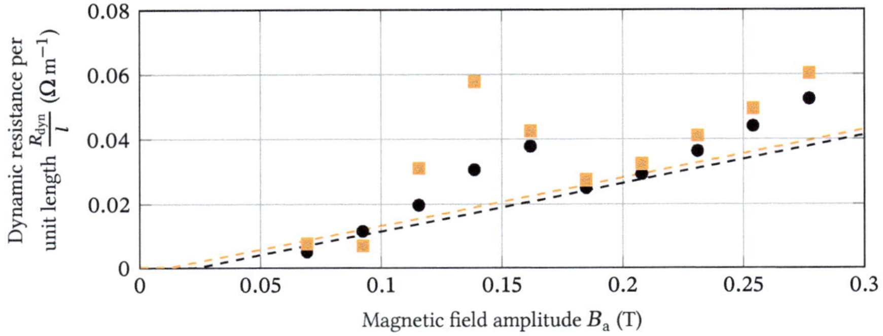

(b) 1000 Hz

Figure 3.16: Dynamic resistance R_{dyn} of config. A without polyimide laminate.

The total resistances were also measured for the modifications presented in section 3.2.3 and shown in fig. 3.17. To prevent damaging the non-stabilized superconducting tapes, only a transport current of 3 A is investigated. To clarify, it is sufficient for this work because later in the switching circuit the dynamic resistance only occurs when the superconducting switch is in off-state with only small leakage currents as transport current. The results are displayed for the frequencies 500 Hz and 1000 Hz. It should be noted that all samples are now laminated with polyimide foil as environmental protection. The results from config. A without lamination are added for comparison.

At 500 Hz, the resistance increases linearly for all configurations. The laminated samples show a faster increase compared to the not laminated config. A. Configurations A and B have a similar curve while config. C show a slight non-linear increase with the highest total resistance of 58.4 $m\Omega\,m^{-1}$ at 277 mT. This behavior is more apparent at 1000 Hz where the non-linear increase is clearly visible. It should be noted that the total resistance of config. A at 277 mT already corresponds to the normal conducting resistance R_{nc} of this configuration.

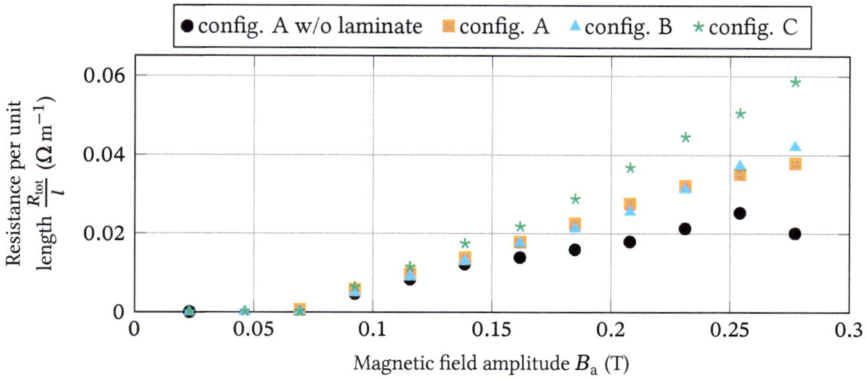

(a) 500 Hz

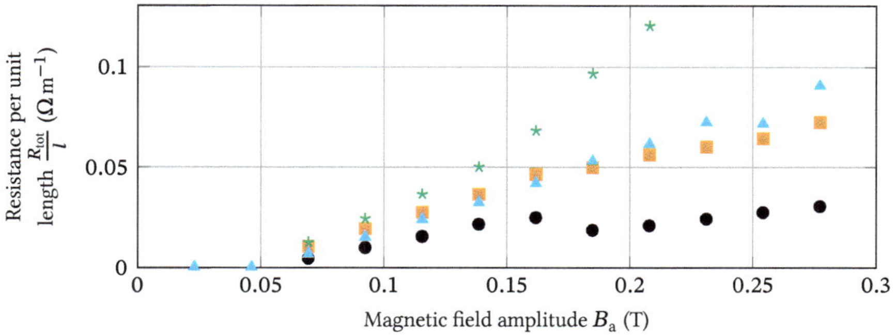

(b) 1000 Hz

Figure 3.17: Measurement of the total resistance R_{tot} of various tape configurations.

The corresponding calculated dynamic resistances are shown in fig. 3.18. For 500 Hz, all configurations show very similar dynamic resistances up to approximately 150 mT while config. A without laminate and config. B having an almost identical progression of the dynamic resistance. The same behavior is also seen at 1000 Hz whereby config. A increases very strongly in a non-linear fashion. It should be noted that the normal conducting resistance R_{nc} for each configuration is considered constant for the calculation of the dynamic resistance. Changes in the normal conducting resistivities, for example due to temperature changes, are not taken into account and have an influence on the actual level of dynamic resistance.

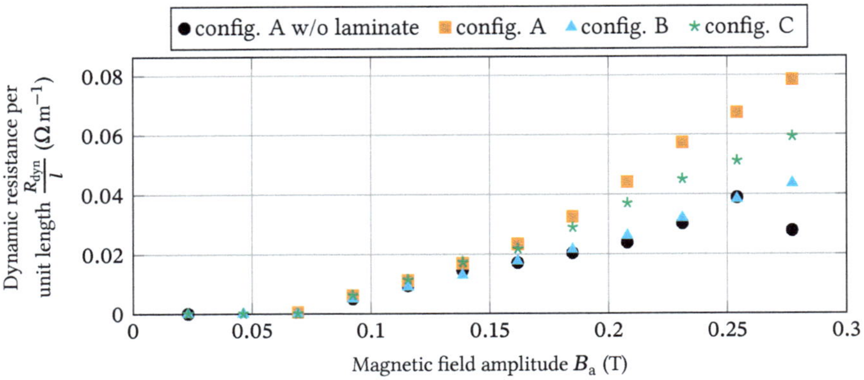

(a) 500 Hz

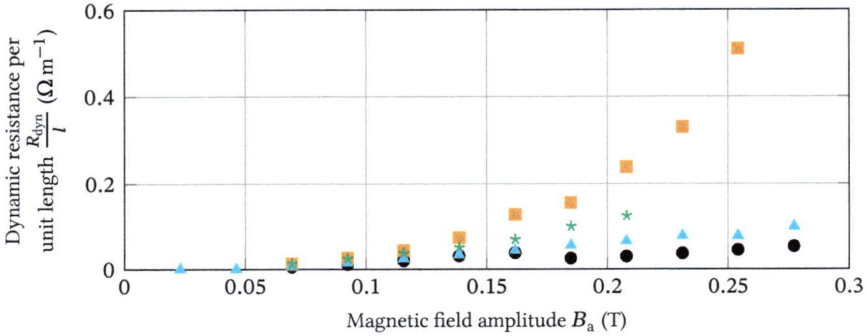

(b) 1000 Hz

Figure 3.18: Dynamic resistance R_{dyn} of various tape configurations.

Since the total resistance in fig. 3.17 is already in the region of the normal conducting resistance R_{nc}, it can be assumed that the boundaries of superconductivity are already exceeded and a temperature rise above the critical temperature T_c is expected.

3.3.2 Comparison with FEM Model

In order to verify the measurements, multiple modeling approaches are investigated. As shown, the measurement data differs strongly from the analytic models. This is expected since these models, as described in section 2.3, are only suitable for low magnetic field and low current load for eq. (2.13) and high magnetic field and high current loads for eq. (2.17). With the measurements made, the case of high magnetic fields at low current load ratio$i$ 0.0088 is examined, and the normal conducting layers cannot be neglected [Zha20b]. Therefore, a numerical modeling of the whole superconducting tape in this particular case is necessary.

The H-formulation presented in section 2.3.4 is used as the modeling method. Since this model also considers the normally conducting layers of a superconductor, it provides a detailed view of the losses and resistances of the individual layers. The modeling is performed in COMSOL Multiphysics® [COM]. In the 2D model, the cross-section of one superconducting tape displayed in fig. 3.19 is simulated using the parameters from table 3.2. The magnetic field dependency of the critical current is modeled by eq. (2.6) with a critical current I_c of 338 A and the model parameters $B_{ch} = 42.65\,\mathrm{mT}$, $k = 0.295\,15$, and $b = 0.7$. Two cases are investigated, first the unmodified tape and second the etched tape without silver. Both neglect the buffer layer due to its high resistance. The results are plotted in figs. 3.20, 3.21 and 3.23.

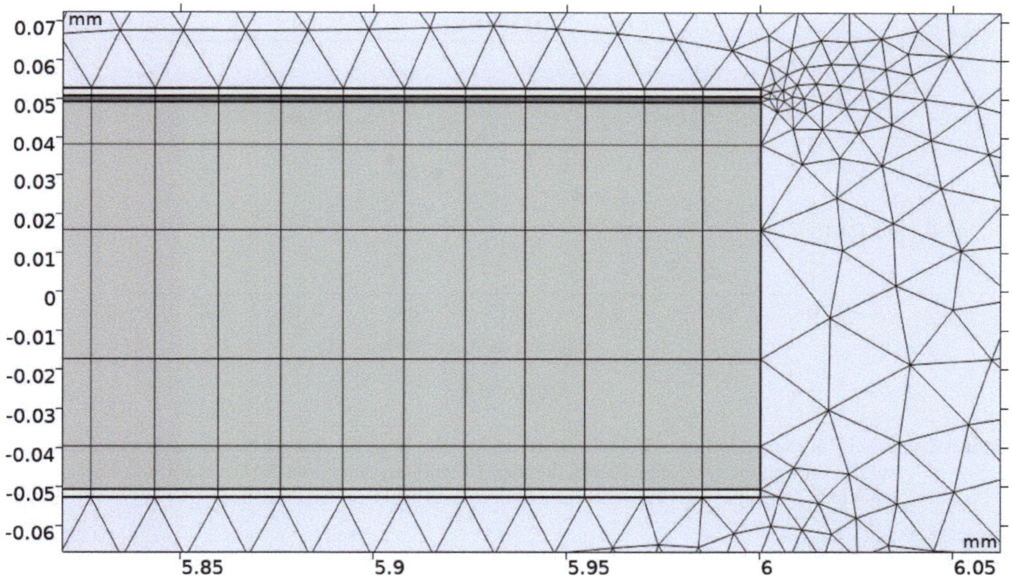

Figure 3.19: Section of the modeled geometry in COMSOL Multiphysics® with all layers of a superconducting tape and the surrounding air. The tape has a mapped mesh and the air is triangular meshed. The displayed domains are air □, substrate ■, superconducting layer ■, and silver □.

Figure 3.20 shows the measured total resistance of config. A without laminate (●) in comparison with the linear analytic eq. (2.13) (——) and the non-linear eq. (2.17) (——) with eq. (2.8) and the numerical results (■). Since the current load ratio is so small, the linear and non-linear equation are identical. At lower magnetic fields the numerical results and the analytic data give similar results but begin to deviate between 150 mT and 200 mT, with the numerical model giving higher resistances. The measurement results are initially higher than the simulation, but after the drop both are very similar.

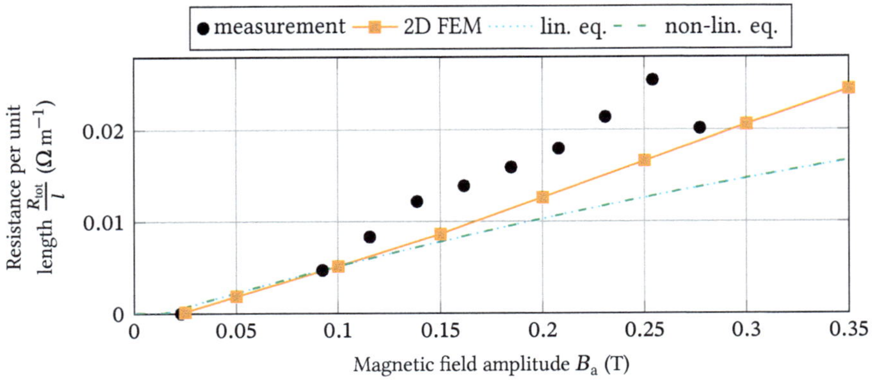

(a) 500 Hz

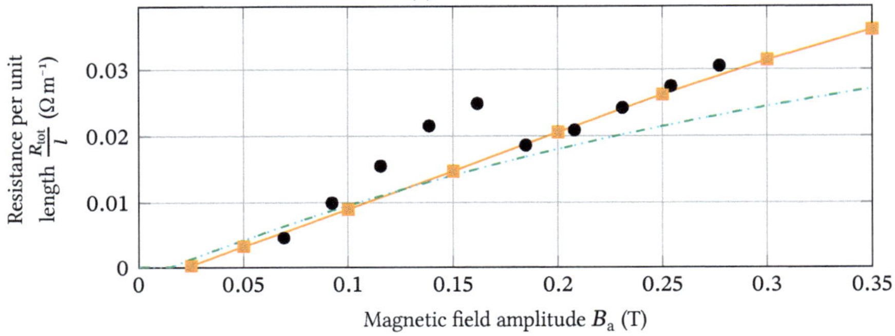

(b) 1000 Hz

Figure 3.20: Results of the numerical model of total resistance R_{tot} with silver layer compared to measurement results of the non-modified superconducting tape in configuration A and the linear and non-linear analytic equations. Displayed are the results for a frequency of 500 Hz and 1000 Hz.

With the numerical simulation it is possible to calculate the dynamic resistance via the transport losses. These results are shown in fig. 3.21. Again, the results are compared with the non-linear equation and the measurement of config. A without laminate. The simulation clearly shows a non-linear progression and agrees well with the measurement after the drop. It also shows that the non-linear equation massively underestimates the dynamic resistance in the case of high magnetic fields and low current loads. Additionally, the results of the simulation without silver is plotted with the same progression as with silver, but with a certain offset higher. This behavior is apparent when the distribution of the transport current to the individual layers is considered. Moreover, the distribution in both cases is displayed in fig. 3.22 for a transport current of 3 A and 1000 Hz as frequency of the external magnetic field. Due to the good conductivity of silver, the transport current is redistributed from the superconducting layer to the silver as soon as the dynamic resistance comes into the region of the silver resistance. At 330 mT, the transport current in the silver is higher than in the superconducting layer and carries half of the transport current. Because of the high resistance, the substrate layer contributes only little to the current transport and carries a maximum of 2.6 % in the simulation with silver and 6.9 % in the simulation without silver. The transport current in the superconducting layer is therefore higher without silver and results in the offset in the dynamic resistance.

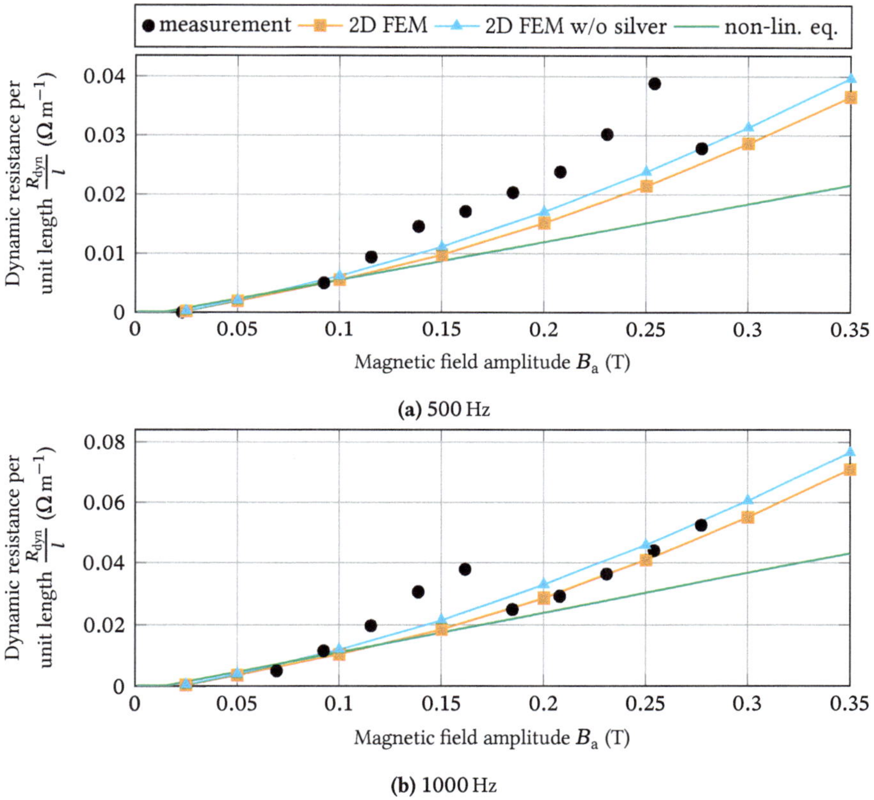

Figure 3.21: Results of the numerical model of dynamic resistance R_{dyn} with and without the silver layer compared to measurement results of the non-modified superconducting tape in configuration A and the non-linear analytic equation. Displayed are the results for a frequency of 500 Hz and 1000 Hz.

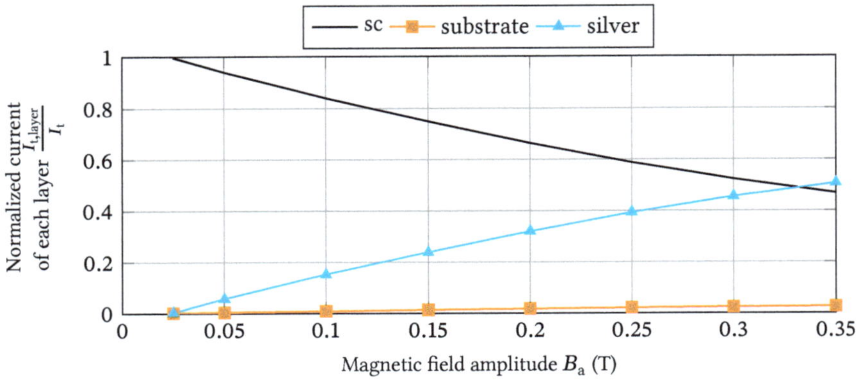

(a) with silver

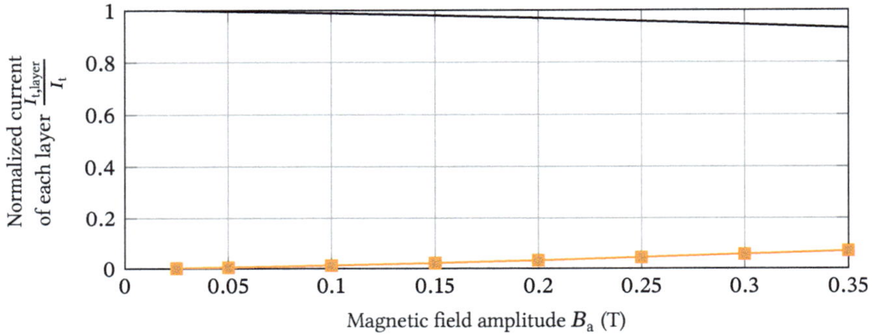

(b) without silver

Figure 3.22: Distribution of the transport current to the individual layers of the superconducting tape from the 2D FEM model. For the case with and without silver layer at 1000 Hz and a transport current of 3 A.

In order to better understand the trend of the measurements, the losses of the individual layers are examined in more detail. These are displayed in figs. 3.23 and 3.24. Figure 3.23 shows the total losses Q_{tot} which combines transport and magnetization losses of each layer at both frequencies. Generally, they increase steadily with the external magnetic field. It can be seen that the substrate contributes little to the total losses of the superconductor. The losses in the superconducting layer increase with a flattening slope and is the largest loss component at 500 Hz and scale linearly with the frequency. The losses in the silver layer are mainly eddy current losses which depend quadratically on the magnetic field amplitude and frequency [Mel24]. At 1000 Hz they increase so steep that at around 330 mT they exceed the losses of the superconducting layer. The removal of the silver layer has little impact on the losses of the remaining layers. As displayed, the total losses without silver are the sum of substrate and superconducting losses, whereby the superconducting losses are the dominating factor.

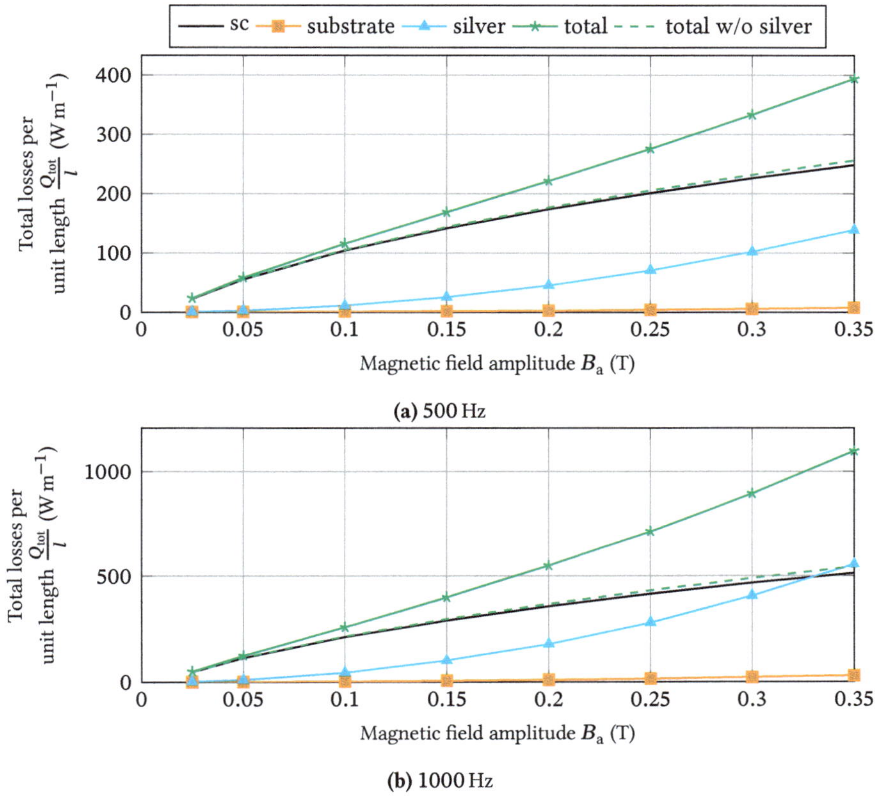

Figure 3.23: Numerical modeling of losses in different layers of a superconducting tape with and without silver layer for 500 Hz and 1000 Hz. Transport current is 3 A.

Figure 3.24 divides the total losses into transport losses Q_{trans} and magnetization losses Q_{mag}, i.e. losses caused by the external alternating magnetic field. The losses of the layers are stacked on top of each other, so that the total transport and magnetization losses are also obtainable. Only the losses at 1000 Hz are plotted. It can be clearly seen that the majority of the losses are caused by the magnetic field, since the transport current of 3 A is small. Looking at the transportation losses, the current distribution from fig. 3.22 is reflected. Both silver and the superconducting layer, which carry most of the transport current, also generate most of the transport losses. However, these are several orders of magnitude lower than the magnetization losses. The distribution of magnetization losses is similar to that of transport losses, and as the magnetic field increases, the losses in the silver layer increase more and more with the magnetic field due to the quadratic dependence.

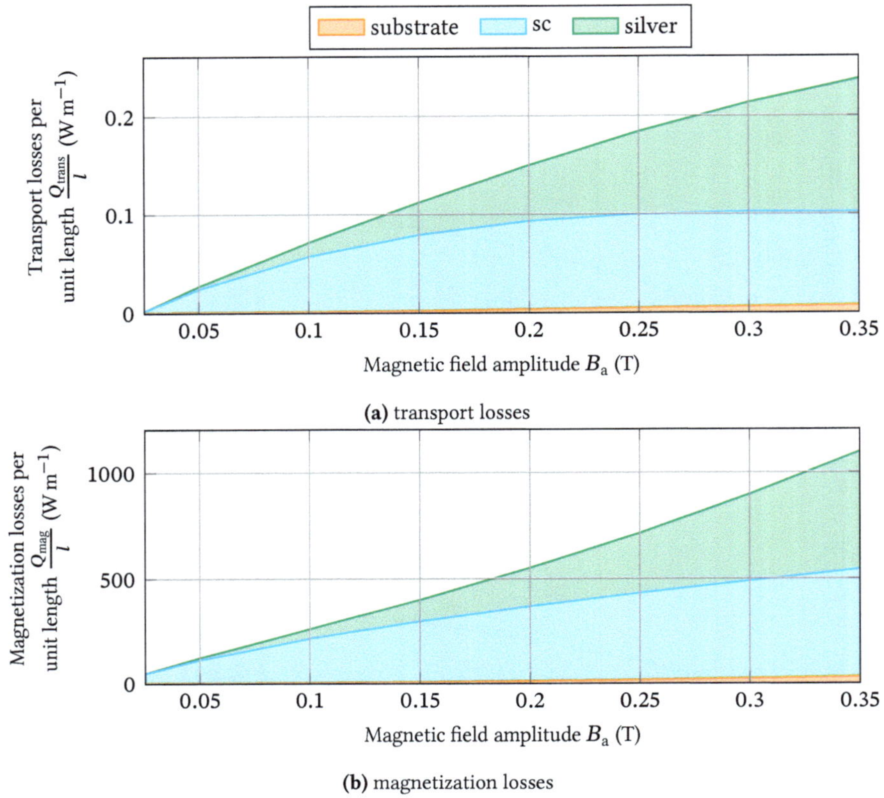

Figure 3.24: Stacked transport losses and losses caused by the magnetic field of the 2D FEM model divided to the layers of the superconducting tape. The frequency is 1000 Hz.

In the measurements, the generated losses are dissipated via pool boiling in LN_2. How effective it is depends on the heat flux density $\dot{q}$ from the superconducting sample to the cooling fluid. This can be calculated from the total losses and the geometry of the superconducting tape and is displayed in fig. 3.25 for the investigated numerical models. The two extreme cases are 500 Hz without silver and 1000 Hz with silver. The former achieves a maximum heat flux density of 1.07 W cm^{-2} and the latter increases the heat flux density up to 4.55 W cm^{-2}.

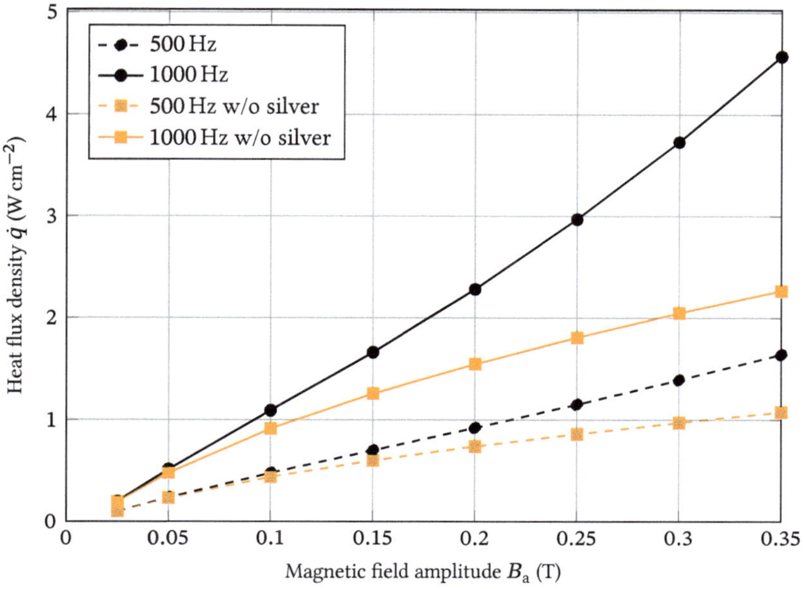

Figure 3.25: Heat flux density of the modeled superconducting tapes.

The obtained heat flux density can then be compared to measured boiling curves of LN_2 from literature [Mer64] where Hellmann et al. [Hel14, Hel19] have examined different surface finishes, including polyimide lamination. From this, the excess temperature ΔT can be derived. Taking the heat flux density of 3 W cm^{-2} at 1000 Hz and 250 mT of the model with silver, this would mean a temperature increase of 4 K to 5 K for an untreated surface and 30 K to 40 K for polyimide laminated. This corresponds well with the achieved total resistance in fig. 3.17 and resistances from the $R(T)$-behavior in fig. 3.13. Due to the high losses of the superconducting tape under external magnetic field, the temperature increases by up to 40 K and highly impacts the achievable total resistance. Most of the losses are generated in the silver and superconducting layer. Removing the silver layer results in lower heat flux density and therefore the cooling of polyimide laminated samples is sufficient, leading to similar dynamic resistances compared to config. A without laminate. However, the dynamic resistance is strongly dependent on the temperature, which depends on the amount of losses and cooling efficiency. Removing silver from the superconducting samples leads to higher total resistances under the same conditions.

3.4 Commutation Behavior

In the previous sections a full characterization of one superconducting switching unit is presented. The magnetic field dependent critical current and the built-up resistance of the switch is measured. Additionally, options for increasing the total resistance are presented. These superconducting switches are now connected in parallel to prove the switching functionality and quantize the commutation behavior of such a circuit.

For the following experiments the test setup in fig. 3.2 is used. This setup allows a parallel connection of two superconducting paths, each equipped with an electromagnet from section 3.1.1 and current transducer, see section 3.1.2. The current is fed in via the outer copper clamp terminals and the voltage drop is measured inside and outside the copper terminals. When connecting two superconductors in parallel the current distribution depends solely on the contact resistance differences of both paths. The superconductor with the lowest contact resistance, i.e. the one closest to the connected copper terminal, carries most, if not all, of the dc current, since the superconductor itself has no resistance [Cha11, Wil15, Liu22]. In order to prevent this, the contact resistance between both branches is removed by only using one 12 mm wide superconducting tape which is partially slit along the length. This results in two ideally connected superconducting branches that are each 6 mm wide. A schematic drawing with the relevant dimensions is shown in fig. 3.26. Whereas, the slit has a length of 23 cm and is done by laser structuring. This process ensures minimal damage to the superconducting tape. The voltage taps are connected directly after the slit with a distance of 23.5 cm and the second pair is connected outside the copper terminals with a distance of 34 cm. As pictured in fig. 3.2, the selected dimensions ensure that there are no sharp bending radii which could lead to a degradation of the critical current [Ott16].

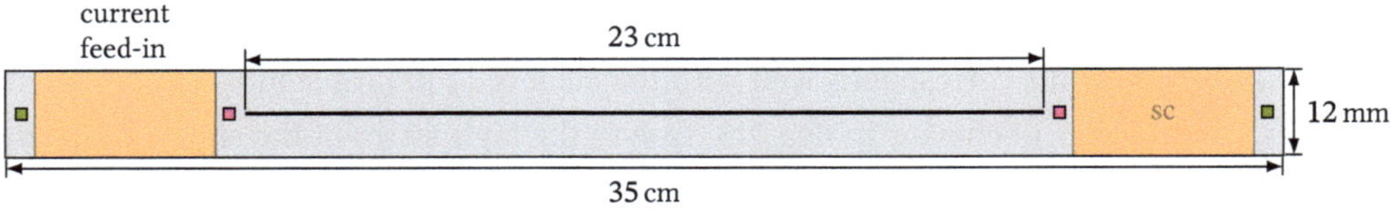

Figure 3.26: Schematic drawing of the slit HTS tape. Both 6 mm wide branches are connected without contact resistance. The areas marked in ■ are the positions of the current feed-in. Also, the inner ■ and the outer ■ voltage taps are displayed.

After laser structuring, the total critical current is measured to ensure that no severe damage to the superconductor occurred and both branches are similar in current carrying capacity. Figure 3.27 shows a schematic drawing of the test setup with the input and output terminals. Since this is a parallel connection, the total current i_0 is the sum of the branch currents i_1 and i_2,

$$i_0 = i_1 + i_2. \tag{3.4}$$

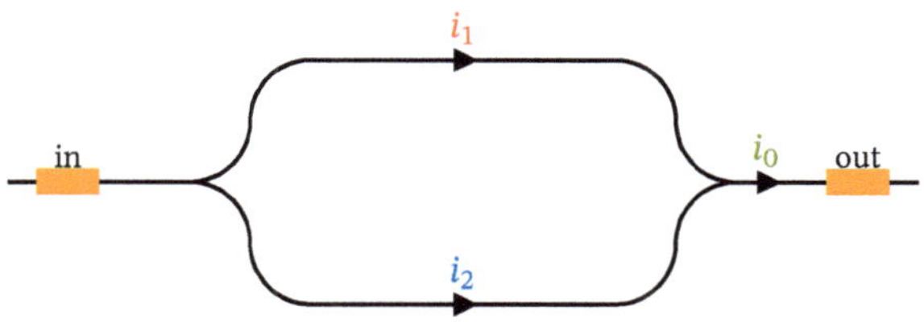

Figure 3.27: Schematic drawing of the test setup for measuring the critical current of the slit superconducting tape.

The results are displayed in fig. 3.28. The top figure shows the expected exponential dependency of electric field E and current I_{tot}, as already described in section 2.1. In this case, the measurement (●) contains a small linear component, which probably comes from the position of the voltage taps too close to the slit. To determine the critical current by fitting the results to the power law, a slightly modified equation as noted in eq. (3.5) is required. Compared to eq. (2.1), this equation adds a linear component with c_1 and an offset parameter c_0 in order to filter out these components and determine the critical current and the n value:

$$E(I) = E_c \cdot \left(\frac{I}{I_c}\right)^n + c_1 \cdot I + c_0 \tag{3.5}$$

The fitted parameters are listed in table 3.4 for the whole tape and each branch, and are displayed in fig. 3.28. The bottom graph shows the current distribution to both branches dependent on the total current. As one can see, the current is almost equally distributed to both parallel branches and both are increasing linearly with the total current. However, at higher currents, branch #2 is carrying slightly more current than branch #1. At a total current i_0 of 250 A the current is divided into i_1 = 123.9 A in branch #1 and i_2 = 126.1 A in branch #2. To determine the critical current of each branch, the same approach as for the whole tape can be utilized, where the electric field obtained from the measured voltage divided by the length of the slit (23 cm) and the input parameter is the branch current. The critical currents of the individual paths are almost identical, they differ by about 2 A and can be found in table 3.4. The linear component shown in fig. 3.28 can be used to determine the resistance of 14.87 nΩ for the inner voltage taps and 9.07 nΩ for the outer voltage taps. These resistances are caused by the position of the voltage taps that are too close to the copper feed-in terminals and therefore are located in the current transfer length [Eki06, p. 291]. The linear component occurring here is in the nΩ-range and can therefore be neglected.

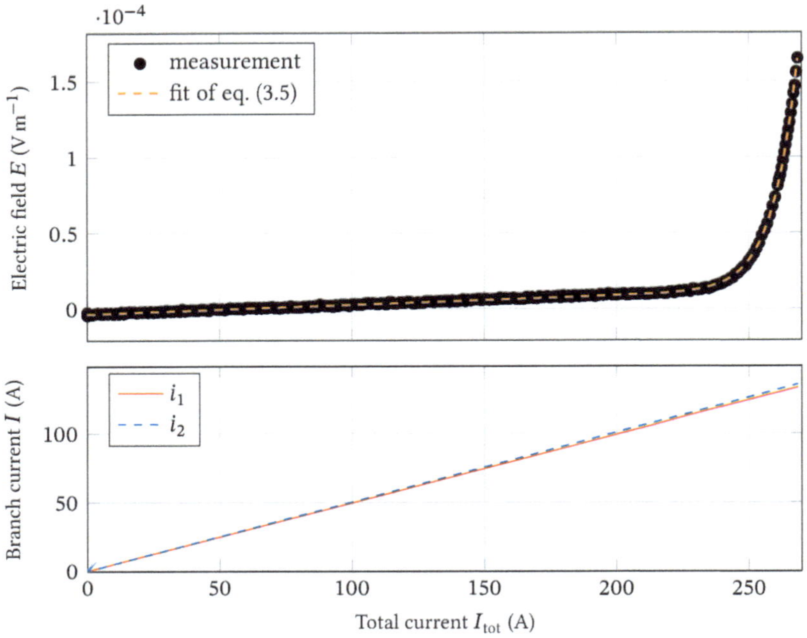

Figure 3.28: Critical current measurement of the slit superconducting tape. The graph above shows the voltage drop per unit length between the inner voltage taps (distance = 23.5 cm) as a function of the total current and the graph below shows the total current divided to the two branches.

Table 3.4: Parameters of curve fitting to obtain the critical current

Inner voltage taps

	I_c (A)	n	c_1 ($\mathrm{V\,A^{-1}\,m^{-1}}$)	c_0 ($\mathrm{V\,m^{-1}}$)
Branch #1	130.90	29.06	$1.25 \cdot 10^{-7}$	$-3.12 \cdot 10^{-6}$
Branch #2	133.10	28.72	$1.25 \cdot 10^{-7}$	$-3.20 \cdot 10^{-6}$
Total	264.41	28.53	$6.22 \cdot 10^{-8}$	$-3.13 \cdot 10^{-6}$

Outer voltage taps

	I_c (A)	n	c_1 ($\mathrm{V\,A^{-1}\,m^{-1}}$)	c_0 ($\mathrm{V\,m^{-1}}$)
Branch #1	130.87	28.86	$7.41 \cdot 10^{-8}$	$-2.82 \cdot 10^{-6}$
Branch #2	133.08	28.38	$7.46 \cdot 10^{-8}$	$-2.90 \cdot 10^{-6}$
Total	264.37	28.45	$3.81 \cdot 10^{-8}$	$-3.03 \cdot 10^{-6}$

This test setup provides two parallel superconducting branches that are connected with zero contact resistance. Additionally, the branches are arranged geometrically identical and the inductance of the branches are therefore the same. The current is distributed equally to both branches and the setup is suited to investigate commutation processes between two superconducting paths under different external magnetic fields.

3.4.1 Single-Sided Triggering

To see whether the current can be influenced or controlled at all with the dynamic resistance, one of the parallel superconducting branches is equipped with an electromagnet. By switching the magnetic field on, the current is in theory fully redistributed to the other branch. In fig. 3.29 a schematic drawing of the test setup with one mounted electromagnet that generates the magnetic field $B1$ is displayed.

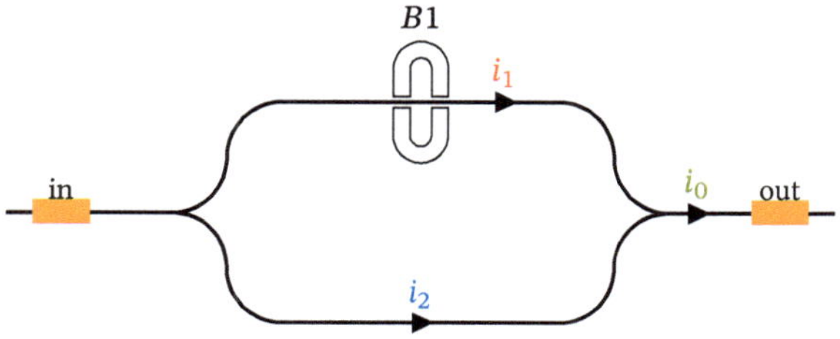

Figure 3.29: Schematic drawing of the parallel test setup equipped with one magnet to show the ability of current redistribution.

The electromagnet allows the frequency and amplitude to be adjusted as required within the limits of the power supply. As an example fig. 3.30 shows the results for a frequency of 300 Hz and a magnetic field amplitude of 84.42 mT at a total current of 50 A. The upper graph shows the magnet control time sequence and below the total current i_0, the current in branch #1 i_1, and the current in branch #2 i_2 over time. Also, the time period in which the magnet is turned on is between 1 s and 3 s. At the beginning, the magnet is off, and the current distribution is similar in both branches namely 18 A and 32 A. The fact that branch #1 carries less current is probably due to the magnet, which has a ferromagnetic iron core and is mounted in proximity to the superconducting tape. This will be investigated in more detail in the next section. At $t = 1$ s the magnet is turned on for a duration of 2 s. Thus, the current in branch #1 is decreasing exponentially until the current is fully redistributed to branch #2. In order to define this commutation, the rise time t_r and fall time t_f are analyzed. This is defined as the time needed for the current to change from 10 % to 90 % of its end value. Details on how to obtain the rise and fall time can be found in section C.1. Here the rise time is 210.2 ms. Since this is an exponential process, it is more common to give the time constant which is 95.67 ms according to eq. (C.4).

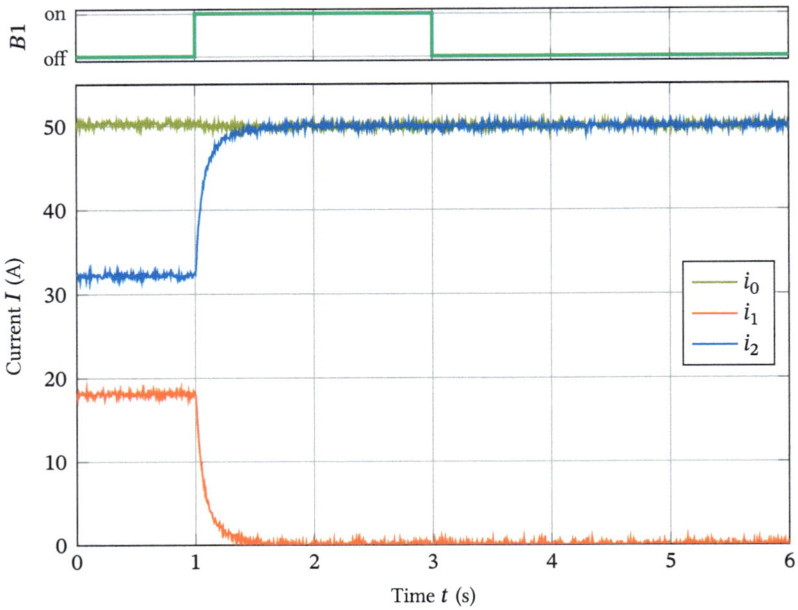

Figure 3.30: Progression of current in the two parallel branches when the magnet is triggered. The total current is 50 A, and the magnetic field has an amplitude of 82.42 mT at a frequency of 300 Hz.

At $t = 3$ s, the magnet is switched off, however, the current does not return to its original distribution. The applied dc current is well below the critical current of one branch, therefore no resistance occurs. This state is intrinsically stable and only changes when it is disturbed externally, e.g. disconnecting the current source or heating of the superconductor.

Concluding, this test shows that by applying an alternating magnetic field and therefore generating a dynamic resistance, a current distribution change of two parallel superconducting branches is possible. The current in the branch with the applied magnetic field can even be reduced to zero.

3.4.2 Double-Sided Triggering

The circuit of an H-bridge inverter from fig. 1.1 consists of two times two parallel connected superconducting switches. This is further investigated by extending the previous test setup with another electromagnet as pictured in fig. 3.31.

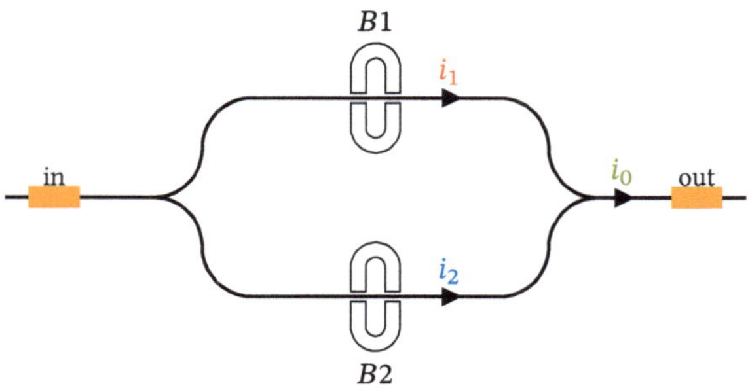

Figure 3.31: Schematic drawing of the parallel test setup equipped with two magnets representing two parallel connected superconducting switches.

The magnets of the switches, each equipped with a separate amplifier, are individually controlled by the four channel waveform generator in an alternating pattern and forces the current to redistribute between the superconducting branches. The trigger sequence is displayed in fig. 3.32 top for both magnets. At first, both magnets are off until 1 s where the magnetic field $B1$ is turned on effectively bringing this switch in off-state. At a time of 3 s, magnet $B1$ is switched off and $B2$ switched on. This is the point in time at which the commutation is later examined more closely and the rise and fall time are determined. After 2 s magnet $B2$ is switched off and the initial state is re-established.

Afterward, the influence of the amplitude and frequency of the external magnetic field on the rise and fall time is investigated by varying both parameters. It is done automatically by a script, which initially increases the transport current to 50 A and then cycles through various parameter combinations. In fig. 3.32 the currents over time are presented exemplary for the set point 100 Hz and 112.56 mT. As expected, the current is redistributing between the two superconducting switches according to the magnet trigger pattern displayed above. The current of the switch in off-state is suppressed to zero while the switch in on-state carries the full current of 50 A. As mentioned before in section 3.4.1, the current commutation between is exponential. After turning off all magnets, the current distribution stays the same as already observed in fig. 3.30. This is also the reason why all the current initially flows in branch #1, since the power supply was not disconnected between parameter set points. The rise time is then evaluated in the area marked with - - -. For this parameter set, it is 111.71 ms and the time constant τ is 50.55 ms according to eq. (C.4).

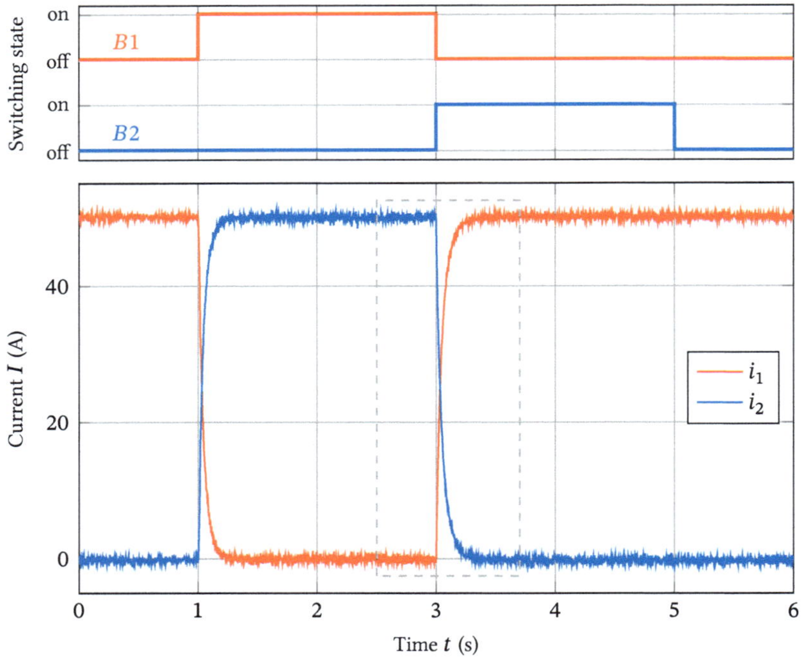

Figure 3.32: Progression of current in the two parallel branches when the magnet is triggered.

Figures 3.33 and 3.34 and table 3.5 show the results of the remaining parameter combinations. In fig. 3.33a, the magnetic field amplitude B_a is varied at a constant frequency of 300 Hz and in fig. 3.33b, the frequency f is varied at a constant amplitude of 225.12 mT to see the impact on the commutation time. Both figures display the current i_2 in branch #2 when magnet $B2$ is switched on at 0 ms. Generally, the higher the field amplitude or frequency, the shorter is the commutation time. The time constant τ decreases from 23.12 ms to 4.18 ms in fig. 3.33a and from 15.67 ms to 4.30 ms in fig. 3.33b. Moreover, a fast Fourier transform (FFT) shows that the current ripple corresponds to the magnetic field frequency. The decrease of the commutation time with increasing frequency or amplitude, is exponential as displayed in fig. 3.34. Table 3.5 summarizes the results of all amplitude and frequency combinations, showing a wide range of time constants from 170.14 ms to 2.52 ms. However, the results also show that due to the exponential decay, much higher magnetic fields and frequencies are required to significantly reduce the commutation time.

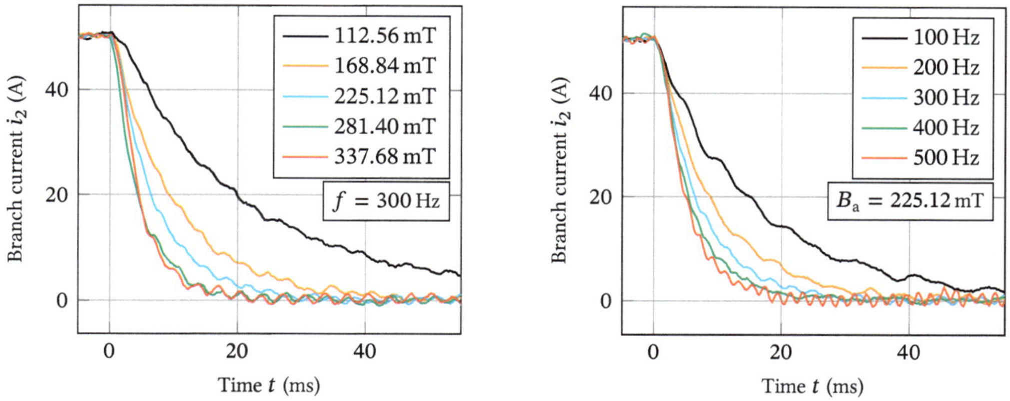

(a) constant f **(b)** constant B_a

Figure 3.33: The progress of branch current i_2 when magnet $B2$ is switched on for different parameter sets.

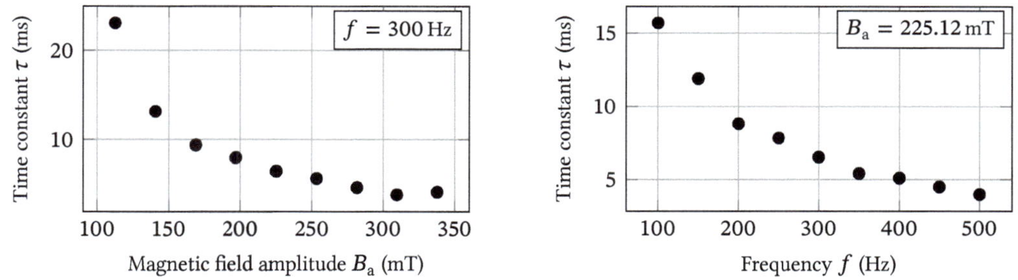

(a) constant f **(b)** constant B_a

Figure 3.34: Progression of the commutation time dependent on the magnetic field amplitude and frequency.

Table 3.5: Commutation time constants for different magnetic field amplitudes and frequencies for double-sided triggering.

τ (ms)	**Frequency f (Hz)**								
Field amplitude B_a (mT)	100	150	200	250	300	350	400	450	500
84.42	170.14	158.35	149.48	145.56	133.60	123.21	124.17	93.95	91.53
112.56	50.55	37.40	30.72	24.68	23.12	19.38	17.25	15.38	13.55
140.70	33.82	23.27	18.93	16.15	13.16	11.49	10.64	9.08	8.57
168.84	24.54	17.53	13.65	11.80	9.45	9.02	7.65	6.56	6.46
196.98	20.56	13.82	11.72	9.41	8.03	7.19	6.42	5.49	4.93
225.12	15.67	11.88	8.84	7.77	6.50	5.43	5.07	4.52	4.03
253.26	13.60	9.14	6.88	6.15	5.71	4.71	4.29	4.04	3.84
281.40	11.12	8.02	6.03	5.68	4.74	4.13	3.63	3.31	3.10
309.54	10.25	6.51	5.68	4.53	3.92	3.47	3.24	2.91	2.89
337.68	7.97	6.24	4.33	4.02	4.18	3.31	3.19	2.55	2.52

In order to link the commutation times with the measured dynamic resistance or total resistance, the equivalent circuit diagram of the test setup is considered which is shown in fig. 3.35. It consists of two parallel branches, each with a variable resistance that is controlled by the magnetic field and the self-inductance of a straight rectangular conductor. The inductance depends on the geometry of the superconducting tape and can be approximated with eq. (3.6) given in [Ros08]

$$L = 2l\left[\log\frac{2l}{G} - 1 + \frac{G}{l}\right] \tag{3.6}$$

where l is the length of the superconducting tape and G the geometric mean distance. For rectangular areas, this geometric mean distance is calculated by

$$\begin{aligned}\log G = \log\sqrt{(2w)^2 + h^2} - \frac{1}{6}\frac{(2w)^2}{h^2}\log\sqrt{1+\frac{h^2}{(2w)^2}} - \frac{1}{6}\frac{h^2}{(2w)^2}\log\sqrt{1+\frac{(2w)^2}{h^2}} \\ + \frac{2}{3}\frac{2w}{h}\tan^{-1}\frac{h}{2w} + \frac{2}{3}\frac{h}{2w}\tan^{-1}\frac{2w}{h} - \frac{25}{12}\end{aligned} \tag{3.7}$$

where $2w$ and h are the width and height of the rectangular shape. This can also be approximated [Gro09, Pią12] by

$$G \approx 0.2235\,(2w + h)\,. \tag{3.8}$$

With the geometric parameters from table 3.2, this gives a self-inductance of approximately 70 nH for each parallel branch. The resistance results from the dynamic resistance and the resistance of the normal conducting layers which is 5.4 mΩ at 300 Hz and 140 mT. The time constant of the RL–circuit which is defined as

$$\tau = \frac{L}{R} \tag{3.9}$$

results in 13.1 ms. This matches well with the data in table 3.5. In order to further decrease the commutation time, either the inductance has to be lowered or the built-up resistance increased.

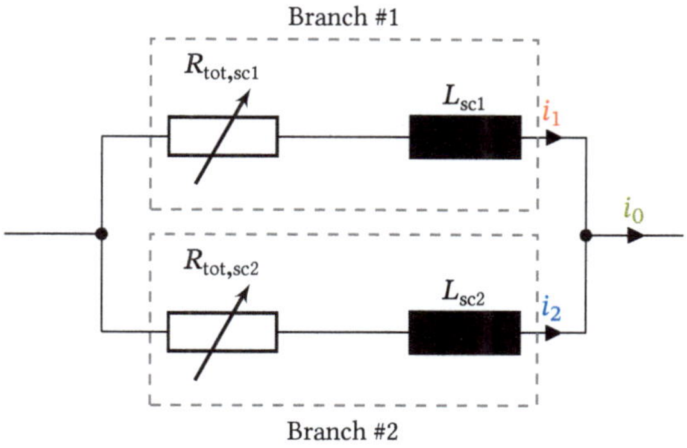

Figure 3.35: Electric equivalent circuit of the experiments with two superconducting branches as displayed in fig. 3.2.

3.5 Summary

In this chapter the relevant parameters of a superconducting switch based on the superconducting tape SF12100 manufactured by SuperPower have been characterized. The magnetic field dependent critical current for perpendicular fields defines the upper limit for the possible current transport of the switch. Since the magnets are switched off in the on-state of the switch, the critical current in relation to the width is $281.67\,\mathrm{A\,cm^{-1}}$ at zero field. The measured temperature dependent resistance represents the upper limit for the achievable resistance when the switch is switched off which is $0.729\,\mathrm{m\Omega\,cm^{-1}}$ at 77 K. In this case, the normal conducting layers, especially the silver cap layer, determine the resistance. In section 3.2.3, methods for increasing the resistance by removing the silver cap layer are presented and an increase by a factor of 40 has been achieved.

The dynamic and total resistances, dependent on the magnetic field amplitude and frequency, are then measured for the different tape configurations, by varying the parameters between 500 Hz and 1000 Hz and from 25 mT to 277 mT at 3 A and 20 A transport current [Pha25a]. The measured dynamic resistances of the different tape configurations are in a similar region since no modifications to the superconducting layer were made and the normal conducting layers have marginal impact on the dynamic resistance. However, adding an environmental protection by polyimide lamination, which also acts as thermal insulation, has a strong influence on the measured resistance.

To verify the measured results a 2D FEM model of a HTS tape with all conducting layers is developed and coincides well with the measurement results. Also, it showed that due to the high ac losses in the superconducting layer and eddy current losses in the silver layer, a significant temperature increase of up to 40 K under pool boiling conditions in LN_2 is possible. This is also reflected in the measured values.

Afterward, the behavior of two superconducting switches connected in parallel is investigated. For that a superconducting tape is partially slit along the length in order to obtain two parallel superconducting branches, each equipped with an electromagnet. This method avoids any influence of contact resistances between the two switches. By switching the superconducting switches in an alternating pattern, a full redistribution of transport current from one to the other branch is achieved. The time it takes to fully redistribute, also called commutation time, is then examined for a large variety of parameter combinations. As a result, a time constant of 2.5 ms was achieved here. These commutation times have been further investigated [Pha22, Pha25b] by modeling a simple RL–circuit with the self-inductance of a thin straight tape. The time constants of the RL–circuit correspond to the measured commutation times.

4 Bridge Inverter Circuit

Following the complete characterization of the superconducting switch and initial investigations into the switching behavior of two switches connected in parallel in chapter 3, the bridge inverter presented in chapter 1 can now be designed and assembled. The circuit is displayed in fig. 4.1.

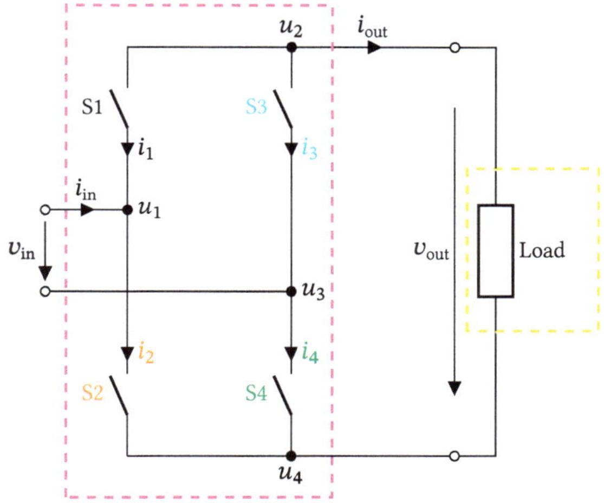

Figure 4.1: Equivalent circuit diagram of a H-bridge power inverter with a resistive load.

Section 4.1 deals with the general circuit and its operation. Furthermore, the three states of switch pattern are introduced and their sequence and the corresponding input and output current of the circuit are discussed. In section 4.2, the implementation with superconducting switches is then explained in more detail, using the same approach to reduce the contact resistance like in the parallel switching experiments from section 3.4. Section 4.3 presents the results of the experiments conducted on the fully superconducting bridge inverter. Individual switching operations are first conducted to show the proper functioning of the circuit. Methods for further reducing the commutation time are also examined in more detail and finally, the stability of the circuit at longer running times is investigated by varying the switching frequency. In section 4.4, models for the full bridge inverter are presented. First, a SPICE model for transients is introduced which can simulate the current distribution between the switches.

Second, a method for efficiency calculation is derived and scaling laws towards higher currents and voltages presented. Afterward, the results are applied to a 10 V/10 kA demonstrator in section 4.5 and various example configurations are calculated. Initial methods for saving superconductor lengths by using wider HTS tapes or sapphire substrate are also presented. Additionally, a method for generating long length magnetic field is demonstrated including experimental investigations.

4.1 Circuit Layout

The H-bridge inverter, also known as single-phase bridge inverter, consists of four switches S1 to S4 which are arranged according to the schematic in fig. 4.1 as the letter "H" with the input as the crossbar. Afterward, the switches are controlled in such way that the voltage and the current is alternating at the output. In this case, the load is assumed to be a resistive load. The present work utilizes superconducting switches based on the dynamic resistance of HTS tapes. The off-state of the switch is achieved by applying an alternating external magnetic field resulting in a dynamic resistance which has already been examined in section 3.3. The on-state is reached by turning the magnetic field off, therefore the superconducting state with zero dc resistance is achieved. The three switching states of the four switches that are relevant in the superconducting bridge circuit are shown in fig. 4.2 including the respective current paths. The resulting output current with the respective switching states is displayed in fig. 4.3.

Figure 4.2a shows the first state where the magnets of the superconducting switches are turned off and therefore all switches are in on-state. This is the default state. The input is short-circuited via the path S1–S3 and S2–S4. Both paths are low resistance, therefore the output current i_{out} is zero. In the second state displayed in fig. 4.2b, switches S1 and S4 are closed and S2 and S3 open. This applies a positive voltage and current across the output. Ideally, the output current i_{out} is equal to the input current i_{in}, since the closed switch is superconductive. This state can be reversed by opening switches S1 and S4 and closing S2 and S3 as shown in fig. 4.2c. The output current is therefore the negative value of the input current. By applying a dc current of i_0 to the input and alternating between state 2 and 3, a square wave current pattern with the amplitude of i_0 is generated at the output of the circuit as displayed in fig. 4.3.

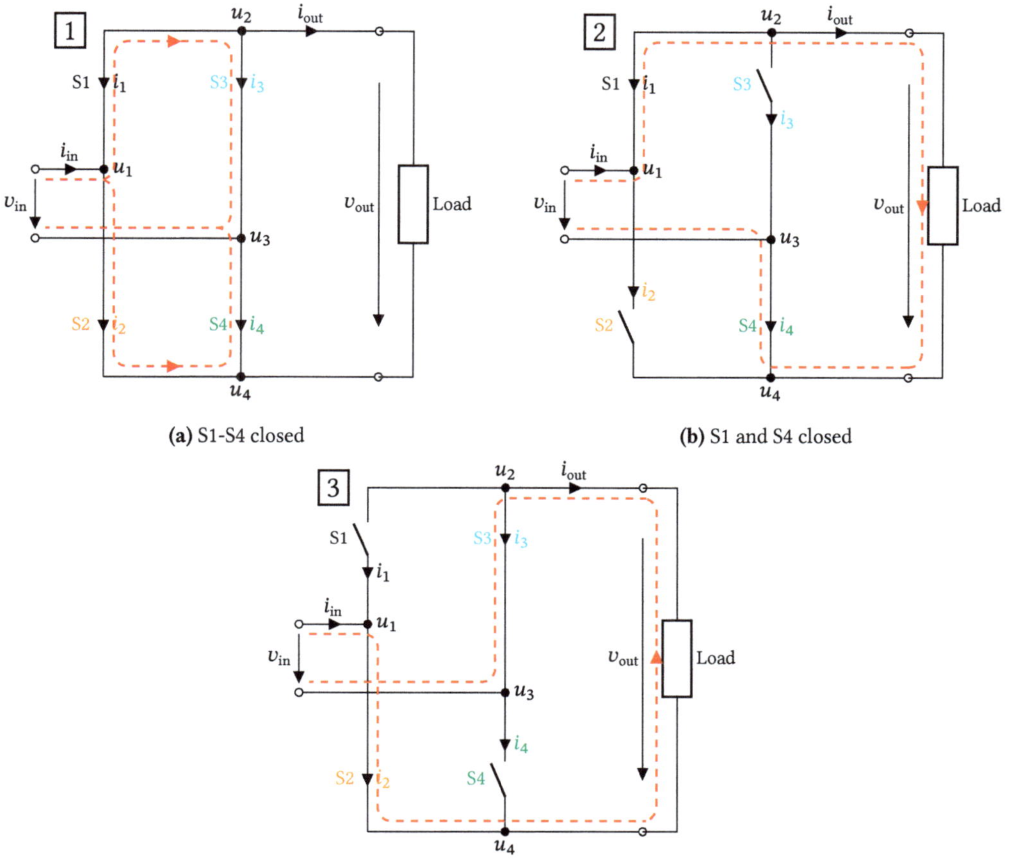

(a) S1-S4 closed

(b) S1 and S4 closed

(c) S2 and S3 closed

Figure 4.2: Relevant switching states of the H-bridge circuit to achieve an alternating output current.

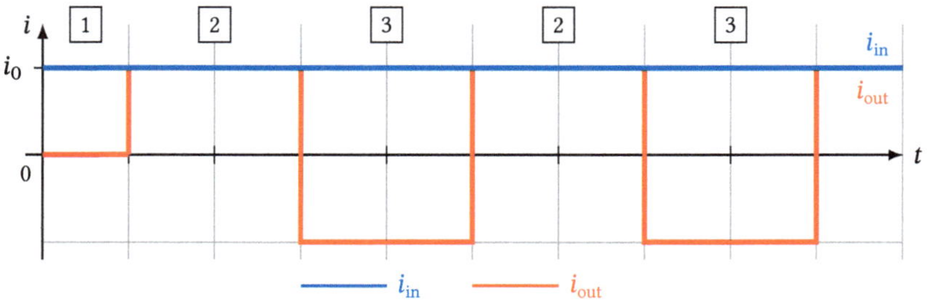

Figure 4.3: Progression of the input and output current of an inverter with the different switching states according to fig. 4.2.

4.2 Circuit Setup with HTS Tape

Figure 4.4 shows a schematic illustration of how such a circuit can be implemented with high-temperature superconductors. It shows four HTS tape sections, each equipped with an electromagnet and arranged to form an H-bridge. If the switches were assembled from individual superconducting tape sections, there would be contact resistances at points u_1 to u_4, which can differ greatly in some cases and thus influence the current flow. Therefore, a similar approach is used as in the setup for investigating the commutation behavior with a slit HTS tape.

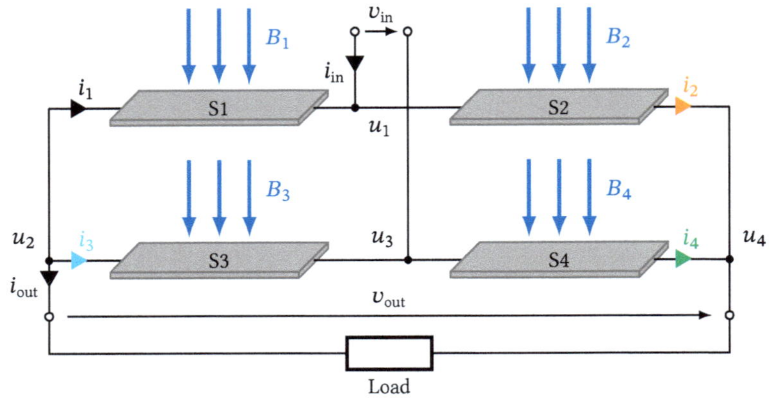

Figure 4.4: Schematic drawing of the H-bridge based on the dynamic resistance of HTS tapes.

The 12 mm wide HTS tape is partially slit along the length as pictured in fig. 4.5 resulting in two superconducting branches connected without any contact resistance. Each of these branches contains two superconducting switches which are also connected without contact resistance as indicated with S1 to S4 in the drawing. The points u_1 to u_4 are marked according to the circuit in figs. 4.1 and 4.4. Moreover, the input terminals at u_1 and u_3 and the output terminals at u_2 and u_4 are realized with copper clamp contacts. Detailed parameters of the HTS tape can be found in table 3.2.

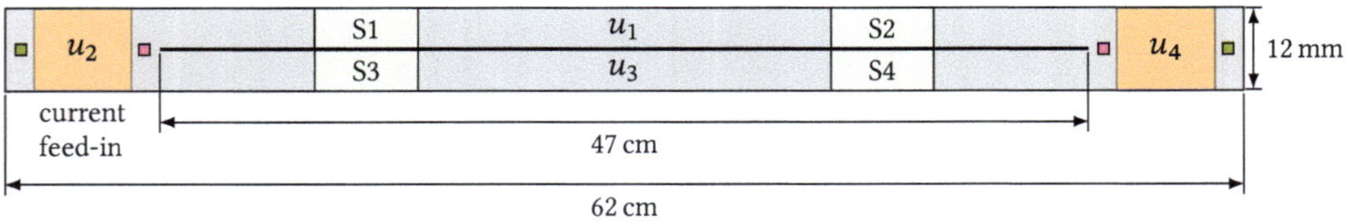

Figure 4.5: Schematic drawing of the slit superconducting tape used for the bridge circuit with a total length of 62 cm and a slit of 47 cm. Detailed parameters of the utilized tape can be found in table 3.2. The areas marked in ■ are the positions of the current feed-in. Also, the inner ■ and the outer ■ voltage taps are displayed.

Figure 4.6 shows the test setup for measuring the critical current of the slit superconducting sample. Each superconducting branch is equipped with a current transducer from section 3.1.2. Furthermore, the critical current is measured across the whole length between the outer copper terminals where inner and outer voltage taps can be found. The distance of the inner voltage taps is 48 cm and 60 cm for the outer. The copper terminals in the middle of the test setup are for the fixation of the superconducting tape in this experiment. The mounting points for the electromagnets of the four switches from section 3.1.1 can also be seen. The critical current is measured without and with disconnected electromagnets which later generate the magnetic field over a length of 2 cm. The measurement results are exemplarily plotted for the case without mounted magnets in fig. 4.7. On the one hand, the top graph shows the measurement of the electric field over the total current and displays the expected progression of the power law with a total critical current of 409.7 A and an n-value of 36.86. On the other hand, the lower graph displays the current distribution to both parallel branches whereby both currents have a similar gradient over the entire range. It can also be seen that both graphs do not start at zero ampere. At a total current of zero ampere, a current of 3.48 A flows in branch #1 and a current of −3.48 A in branch #2. The measurement shown here is the second of four critical current measurements that were carried out in succession. The measurement setup is not warmed up again between tests. A cyclic current is established after the current source from the preceding test is switched off. Due to the lack of ohmic resistance between the two parallel paths, this current does not decay.

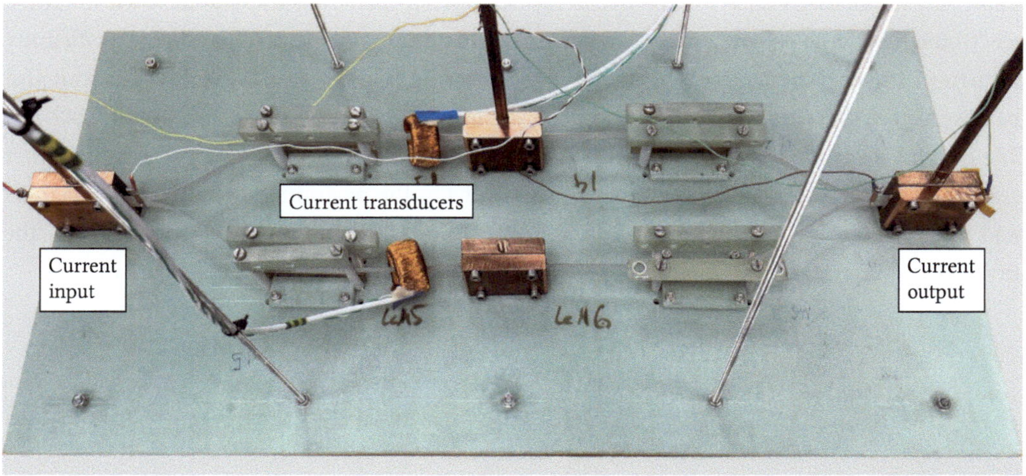

Figure 4.6: Image of the rectifier setup for measuring the critical current. The current is fed in at the two outer copper terminals. Each parallel path is equipped with a current transducer.

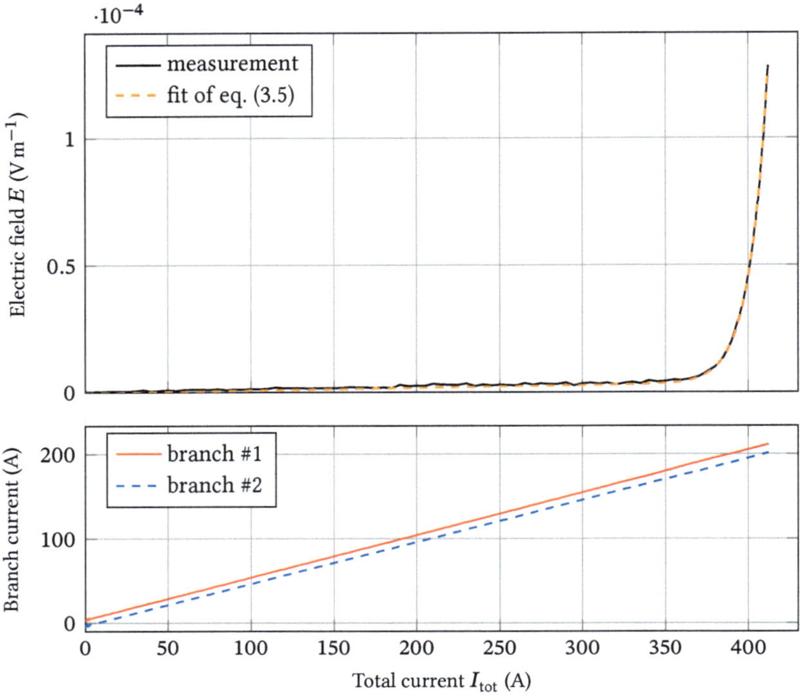

Figure 4.7: Critical current measurement of the slit superconducting tape for the inverter circuit.

Table 4.1 summarizes all critical current measurements including the results with mounted electromagnets. The proximity of a ferromagnetic material to the superconductor significantly influences the current carrying capacity by decreasing it [Göm08, San10, Str19]. This can also be observed in this case where the critical current is lowered by approximately 40 % to 249.1 A. The current distribution also changed slightly with branch #2 now having a marginally higher critical current. This is likely due to small variations in the positioning of the electromagnets and therefore different in the air gap lengths. With a minimal critical current of 120 A, the inverter can be operated with an input current of 50 A with a reasonable safety margin.

Table 4.1: Parameters of curve fitting to obtain the critical current of the slit HTS tape for the bridge circuit.

Without magnets				
	I_c (A)	n	c_1 ($\mathrm{V\,A^{-1}\,m^{-1}}$)	c_0 ($\mathrm{V\,m^{-1}}$)
Branch #1	210.04	38.27	$1.97 \cdot 10^{-8}$	$2.47 \cdot 10^{-7}$
Branch #2	200.05	35.12	$2.00 \cdot 10^{-8}$	$3.60 \cdot 10^{-7}$
Total	409.65	36.86	$9.94 \cdot 10^{-9}$	$3.14 \cdot 10^{-7}$

With magnets				
	I_c (A)	n	c_1 ($\mathrm{V\,A^{-1}\,m^{-1}}$)	c_0 ($\mathrm{V\,m^{-1}}$)
Branch #1	122.81	28.11	$7.60 \cdot 10^{-8}$	$7.04 \cdot 10^{-7}$
Branch #2	126.57	30.71	$7.87 \cdot 10^{-8}$	$1.56 \cdot 10^{-6}$
Total	249.14	29.41	$3.90 \cdot 10^{-8}$	$1.16 \cdot 10^{-6}$

Furthermore, the H-bridge inverter shown in fig. 4.8 is set up. The setup consists of the slit tape, four electromagnets for the superconducting switches, and four current transducers to measure the currents i_1 to i_4 of each switch according to fig. 4.1. The input source is now connected to the copper terminals in the middle. Another 12 mm wide HTS tape, insulated with polyimide tape, is connecting the outer output terminals via face-to-face contact with the slit superconductor acting as a load for the circuit. Voltage taps are placed at both ends of the slit and near the input terminal. The slit superconductor with the four switches and the load tape are highlighted according to the colors in fig. 4.1.

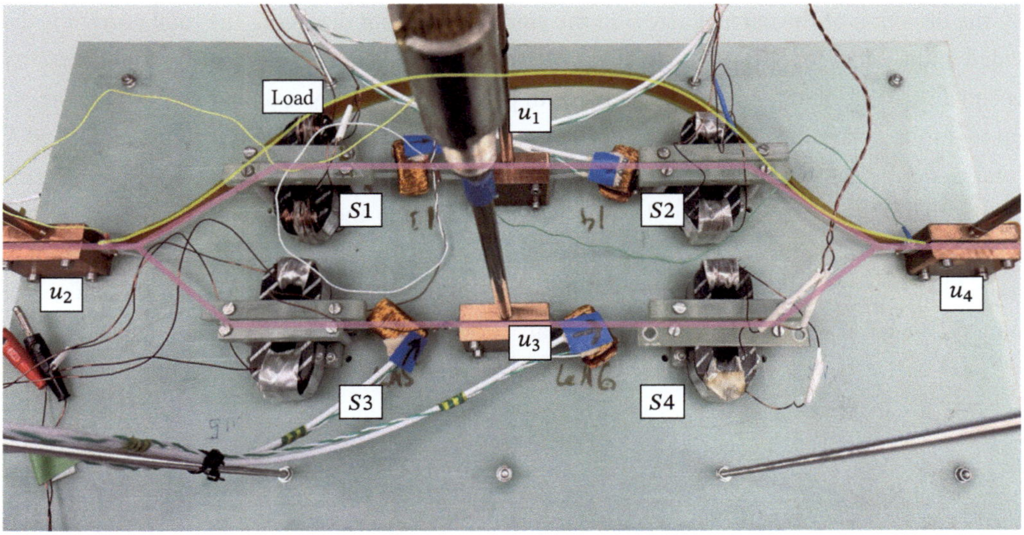

Figure 4.8: Image of the fully set up bridge inverter with HTS tape. The major components, i.e. the slit tape with the four switches and the load tape, are highlighted according to the colors in fig. 4.1.

The measuring and control instruments are similar to the previous experiments. The electromagnets are individually driven by four quadrants power supplies Kepco BOP 72-6 and controlled by an arbitrary waveform generator Tabor WW5064. Additionally, the sampling rate of the digital data acquisition system NI USB-6281 is set to a sampling rate of 10 kS/s for each channel. Also, the power supply Keysight RP4943A is used as input source. The acquisition and controlling is done via a Python script.

4.3 Experimental Results

With the bridge inverter setup, presented in section 4.2, first experiments are conducted in the following.

4.3.1 Direct Current Distribution

The test circuit is submerged in LN_2, and only the input power supply is connected to the circuit in order to measure the current distribution when the input terminals are connected. The electromagnets are mounted but not connected to the power supply. This corresponds to state 1 in figs. 4.2a and 4.3. In state 1, all switches are closed and thus short-circuit the input. The currents of the switches S1–S4 are plotted over time in fig. 4.9 where the input current is increased in 2.5 A steps until it reaches the desired current of 50 A. Also, the sign of the currents is defined by the direction of the current arrow in the equivalent circuit diagram. The total current is divided to the paths S1–S3 and S2–S4, therefore the absolute value of the corresponding currents is equal $|i_1| = |i_3|$ and $|i_2| = |i_4|$. Initially, both parallel paths carry a similar amount of current up to a total current of 15 A, at which point path S2–S4 starts to carry more current. At 50 A total current, the current is divided into 24.3 A in path S1–S3 and 25.7 A in S2–S4. This also means that all the current solely flows in the slit HTS tape and no current flows through the load. Since there is no contact resistance between both paths, the small difference in current is most likely due to small geometry differences which impacts the inductance of the path, or a non-symmetrical feed in the copper terminal.

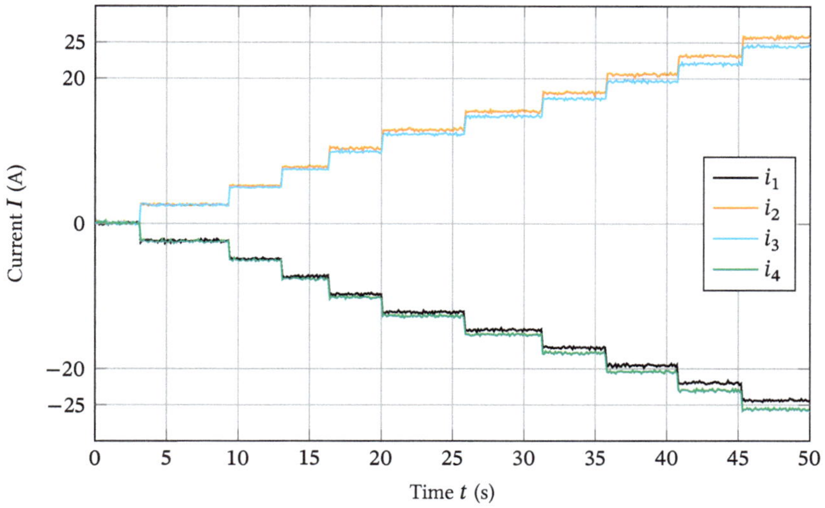

Figure 4.9: Progression of the current in the switches in state 1 of the bridge circuit. The input current is increased to 50 A in 2.5 A steps.

4.3.2 Single Switching Test

Next, the four magnets are connected to their power supplies and first basic switching patterns are conducted to demonstrate the feasibility of a fully superconducting bridge inverter. Plus, the pattern consists of four consecutive states according to fig. 4.2 in the following order: 1–2–3–1. The input current is set to a constant dc current of 50 A and as output a current with a rectangular waveform with an amplitude of 50 A is expected. Furthermore, the same pattern is repeated as already performed in section 3.4.2, for different combinations of frequencies and magnetic field amplitudes to see the impact on the commutation time.

Furthermore, the currents in the switches are displayed exemplarily for a magnetic field amplitude of 169 mT and a frequency of 1500 Hz in fig. 4.10. The top graph shows the switching states of the switches over time. Each switch is only switched off and on once. Below, the current of each switch is displayed. The colors of the line correspond to the colors used in the equivalent circuit from fig. 4.1. In state 1, all switches are turned on and input current is distributed slightly unequal, 30 A over S1–S3 and 20 A over S2–S4. At a time of approximately 5 s, the switches S2 and S3 are opened forcing the current to commutate to the switches S1 and S4, which are carrying nearly the 50 A input current, resulting in switching state 2. Then the switches S1 and S4 are opened and S2 and S3 closed to achieve switching state 3. Again, the input current redistributes to the closed switches suppressing the currents i_1 and i_4 to nearly zero. At approximately 9.5 s, all switches are closed again resulting in a slow redistribution into the initial state. To quantize the current commutation behavior, the rise and fall time between state 2 and 3 is analyzed according to section C.1. In this case, the time from 10 % to 90 % is 6.03 ms.

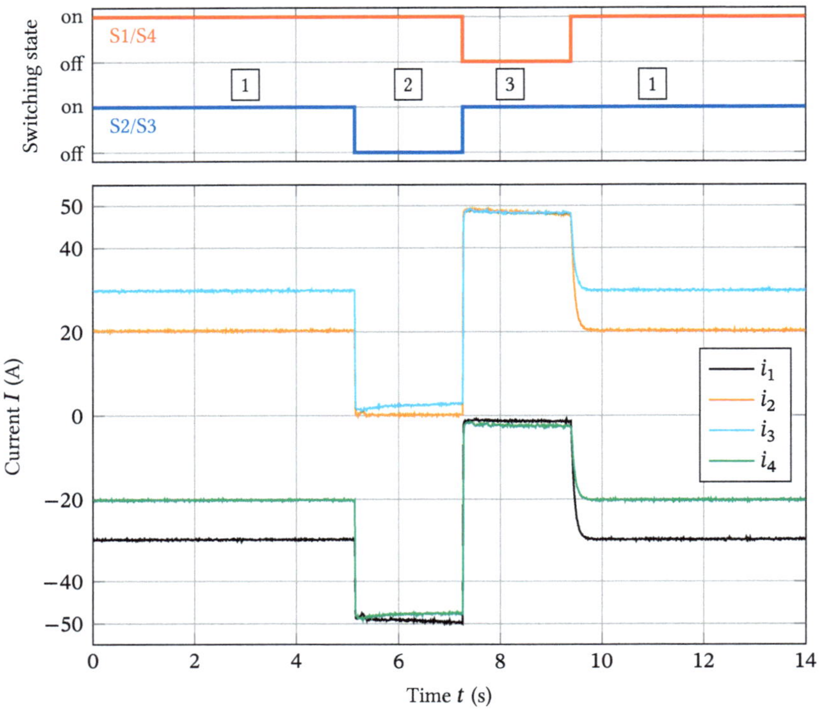

Figure 4.10: Current distribution of the bridge inverter during the basic switching pattern. The switching states of the switches (top) and the current of each switch (bottom) are shown. The electromagnets of the switches are generating a magnetic field amplitude of 169 mT at 1500 Hz.

Figure 4.11 displays the input and output current of the bridge inverter for this example. This is very similar to the current pattern in fig. 4.3. As already mentioned above, the output current does not reach an amplitude of 50 A since a residual current is flowing in the switches in off-state. The so-called leakage currents are approximately 2.6 A in this case. It should be noted that the leakage current is highly dependent on the ratio of the resistance of the switches in off-state and the impedance of the load. This will be further investigated in section 4.4.3.

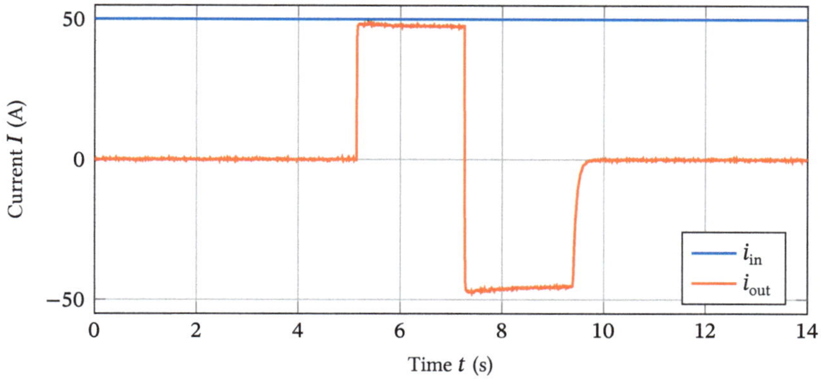

Figure 4.11: Input and output current during the basic switching pattern from fig. 4.10.

4.3.3 Minimizing Commutation Time

To minimize commutation times, the resistance of the open switch plays a major role, as already shown in the chapter on commutation between two parallel paths in section 3.4.2. Various methods of increasing the total resistance generated by an external alternating magnetic field have already been investigated experimentally in section 3.2.3. The different etching patters are now applied to the fully superconducting bridge circuit and some tape configurations, which are displayed in fig. 3.10, are examined in more detail, namely configuration A, the unprocessed HTS tape, and configuration B, where the silver surrounding is removed in a section of 7 cm. Configuration A is examined with and without polyimide lamination and config. B with polyimide lamination only. The same model of HTS tapes as in the previous chapters is utilized with the parameters in table 3.2 and the temperature dependent resistance in figs. 3.9 and 3.13.

Figure 4.12 shows the slit HTS tape for the bridge circuit in config. B. Moreover, the silver is removed in two 7 cm long sections according to fig. 3.10b where the electromagnets will later be located. The etched sections are then laminated in 50 µm thick polyimide foil which acts as environmental protection. The test setup is the same for all configurations, only the slit tape is exchanged.

Figure 4.12: Photograph of the slit superconducting tape in configuration B used for the bridge circuit. The silver is removed in the 7 cm long sections where later the electromagnets are mounted. Displayed is the face side of the superconducting tape and the ReBCO layer is visible.

With the three variants the magnetic field amplitude and frequency of the electromagnets are varied and the commutation time between state 2 and 3 is investigated. Figure 4.13 displays exemplarily the current in switch S3 during the commutation for all three configurations for the set point 138.57 mT and 500 Hz of the magnets. As displayed, the config. B has the fastest fall time with 20.35 ms followed by configuration A with 22.86 ms. The slowest commutation takes place at configuration A without laminate with 26.03 ms. It is also clearly visible that the leakage current differs greatly. The average of the leakage current in this graph ranges from 6.62 A for config. A w/o laminate over 3.10 A for config. A to 97.39 mA for config. B.

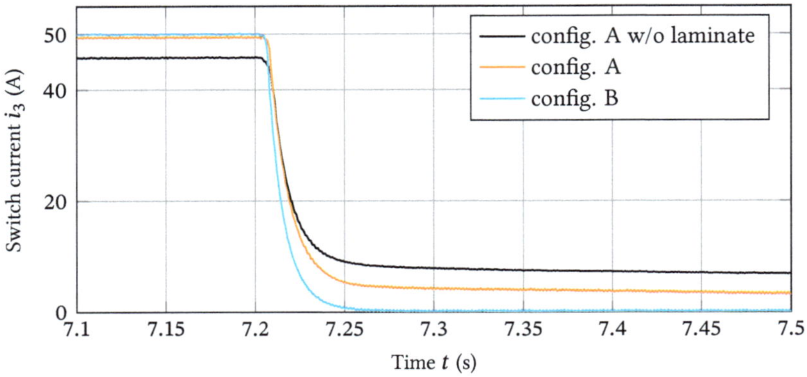

Figure 4.13: Comparison of the commutation behavior of switch S3 in the bridge inverter between different tape configurations. The magnet excitation of the switches is set to 500 Hz and 138.57 mT.

The results of the frequency and amplitude variation are summarized in fig. 4.14, where the commutation times are plotted for 500 Hz (top) and 1000 Hz (bottom) for the different tape configurations. At 500 Hz, commutation times range from 28.9 ms to 6 ms, while at 1000 Hz, they extend from 37.6 ms to 2.32 ms. Configuration A exhibits a gradual decrease in commutation time with increasing magnetic field amplitude but maintains the highest overall values at both frequencies. Configurations with polyimide laminate display nearly identical values and follow a similar decreasing trend. Notably, at high magnetic field strengths, adding polyimide laminate to the HTS tape results in halving the commutation time. These findings align well with the measured total resistance data from section 3.3, where the values of config. A and config. B are remarkably close, mirroring the similarity observed in their commutation times at both frequencies.

These results show a successful reduction of commutation time by removing the highly conductive silver layer and therefore increasing the total resistance.

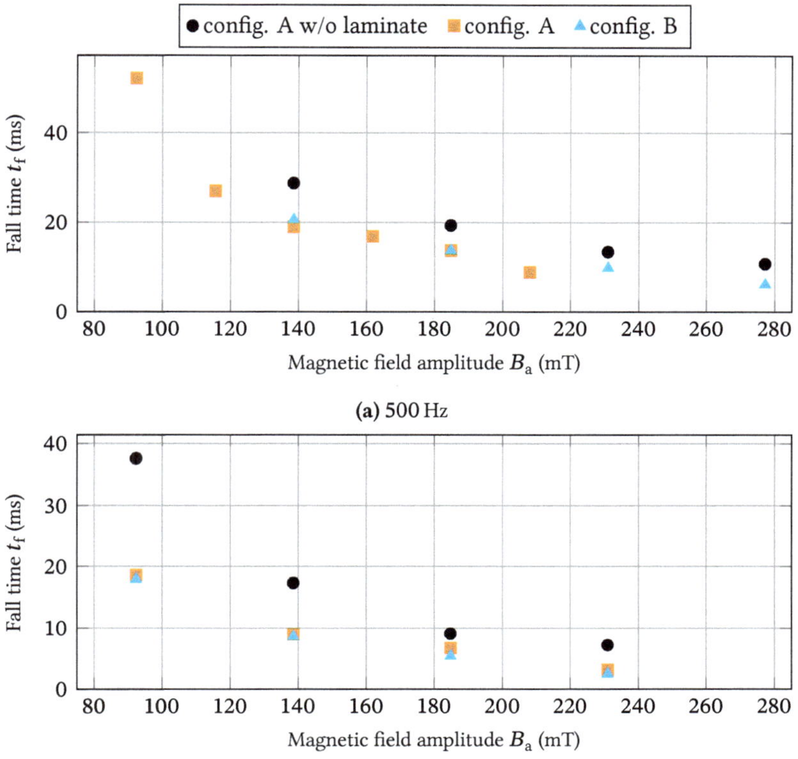

(b) 1000 Hz

Figure 4.14: Comparison of commutation time for different tape configurations at 500 Hz and 1000 Hz under variation of the magnetic field amplitude.

4.3.4 Continuous Switching Test

In general, the switches of an inverter perform multiple switching operations per second over an extended period of time in order to generate the desired waveform at the output. This repeated switching subjects the superconducting switch based on the dynamic resistance to significant electrical and thermal loads. To ensure reliable operation, the HTS tape must maintain its performance without degradation or thermal runaway under these conditions.

The previous experiments were limited to a short-duration test pattern consisting of only single switching pulses for each switch. In order to investigate the scenario above, a continuous stress test is conducted. The test duration t_{on} is varied between 20 s and 120 s and the switching frequency between 1 Hz and 40 Hz which result in pulse widths t_{sw} between 1 s and 25 ms. Additionally, the set point of the electromagnets is altered from 46 mT to 233 mT at two different frequencies, 500 Hz and 1000 Hz, to investigate the impact of the external magnetic field on the stability.

Figure 4.15 (bottom) schematically displays the input and output current of the continuous stress test with the corresponding switching states of the four superconducting switches S1–S4 (top). A dc current with an amplitude of i_0 is applied to the input and the desired output is a square wave with an amplitude of i_0 and a period T_{sq} of $2 \cdot t_{sw}$. The switching state of the bridge circuit is exemplarily numbered for the first three states according to fig. 4.2. The test duration t_{on} is the time in which the switches are actively controlled and the pulse width t_{sw} is the interval in which the control signals of the switches are changed. This means that the switching frequency is twice as high as the frequency of the output current.

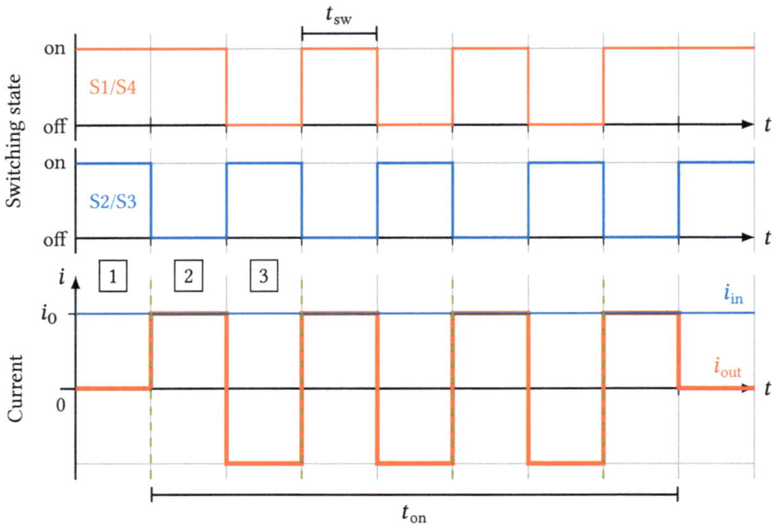

Figure 4.15: Progression of the input and output current of an inverter with the different switching states according to fig. 4.2.

Figure 4.16 shows the currents in the switches i_1 to i_4 over time for various t_{sw} at a constant magnet excitation of 1000 Hz and 138 mT. Each graph displays five switching periods of the continuous test. All show consistent commutation behavior over the length of the test. This was determined by evaluating the commutation times at different points in the test. The commutation times are also very close to each other with the same magnetic field control but variable switching time. The commutation times of the parameters in fig. 4.16 are summarized in table 4.2 and their standard deviation is 0.35 ms. Despite the fast switching times, especially in the 25 ms, the cool down time available between successive commutations remains sufficient. The excess heat in the switches generated by the transport current and the external alternating magnetic field is successfully dissipated and thus enables stable operation of the fully superconducting full-bridge inverter under pool boiling conditions. In addition, at 1000 Hz and 138 mT the leakage current is only about 0.4 A, corresponding to a square wave current amplitude of 49.6 A at the circuit output.

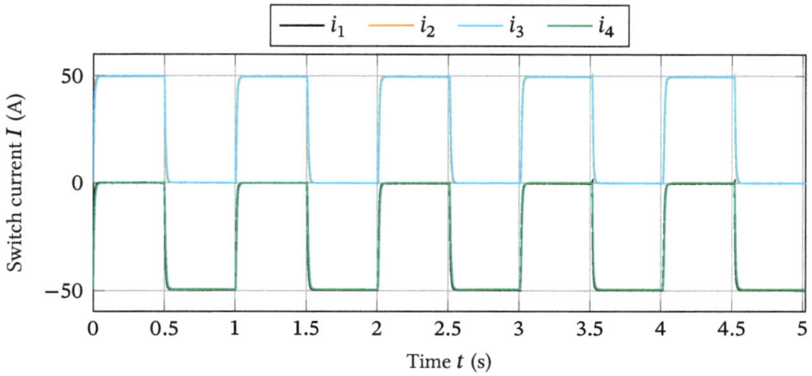

(a) $t_{sw} = 0.5$ s

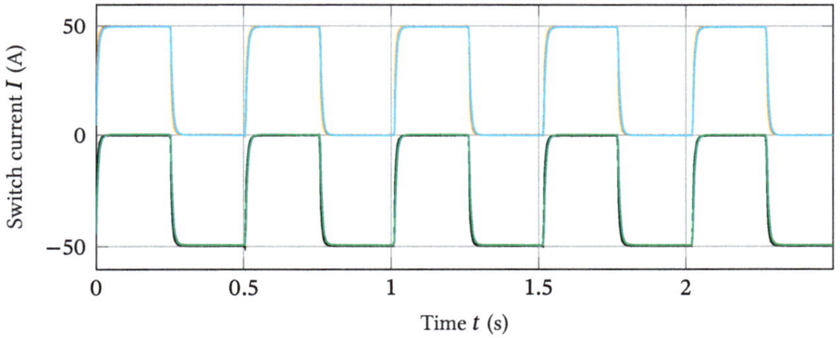

(b) $t_{sw} = 0.25$ s

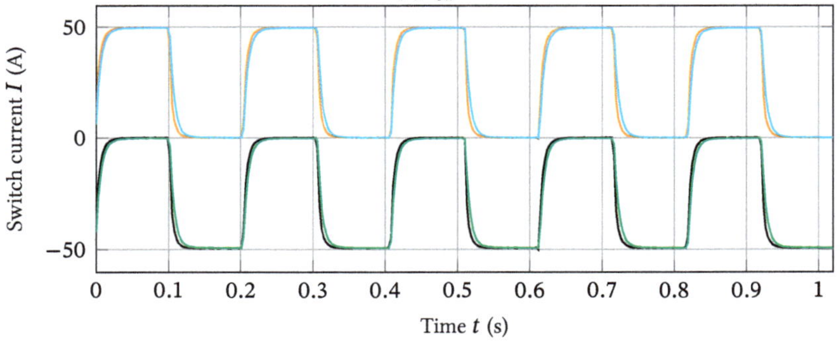

(c) $t_{sw} = 0.1$ s

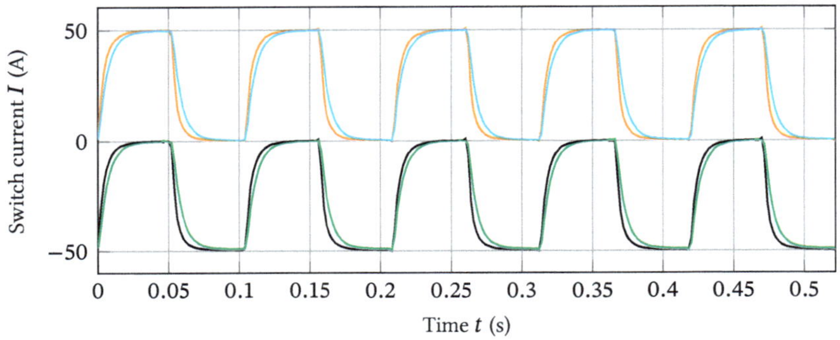

(d) t_{sw} = 50 ms

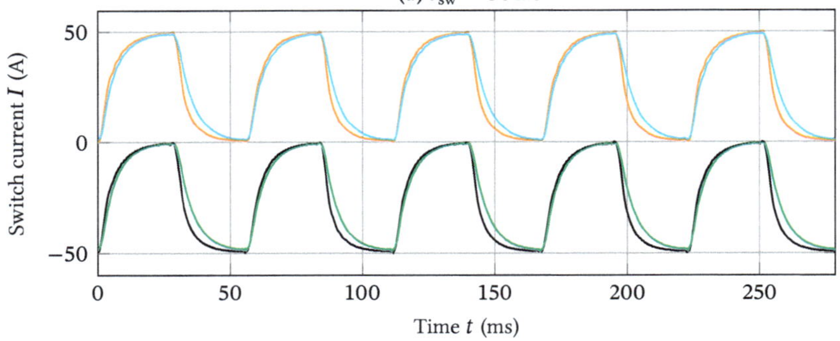

(e) t_{sw} = 25 ms

Figure 4.16: Current distribution of the switches in the superconducting bridge inverter during continuous stress test. The electromagnet set point is 1000 Hz and 138 mT while the switching time t_{sw} is varied.

Table 4.2: Commutation times at a magnetic excitation of 1000 Hz and 138 mT for different switching times as displayed in fig. 4.16.

t_{sw} (ms)	Rise time t_r (ms)	Time constant τ (ms)
0.5	10.79	4.91
0.25	10.53	4.80
0.1	11.37	5.17
0.05	11.30	5.14
0.025	10.98	5.00

Particularly with the shorter switching times, it can be seen that the pulse width t_{sw} is longer than specified. The FFT analysis of the measurement data at t_{sw} = 50 ms results in a frequency of 9.5235 Hz, which corresponds to an average t_{sw} of 52.50 ms, so there is a delay of 2.50 ms. At a pulse width t_{sw} = 25 ms, this results in a frequency of 17.5044 Hz and accordingly means a switching time of 28.56 ms, i.e. a delay of 3.56 ms. This delay is caused by the addition of the individual times for processing the trigger command and running the control program on a non-real-time operating system. The switching time is handled by the function `time.sleep()` in Python 3.12.4 which has a resolution of 100 ns [Pyt]. Afterward, the command is sent via the software backend virtual instrument software architecture (VISA) (1 ms to 10 ms) [Roh] and General Purpose Interface Bus (GPIB) (30 µs) [NI] to the arbitrary waveform generator which has a trigger delay in the range of 200 ns [Tab]. The bipolar power supplies have a slew rate of $0.4\,\mathrm{A\,\mu s^{-1}}$, which adds approximately 15 µs to get to the maximum output current of 6 A. These delays add up to approximately 2 ms to 11 ms which is in range of the measured systematic delay.

Figure 4.17 shows exemplarily the input and output parameters during the continuous test, with the magnet parameters 185 mT, 1000 Hz, and a pulse width t_{sw} of 0.1 s. As before, the load is a superconducting tape. In fig. 4.17 (a) the input and output currents are plotted, where i_{in} is a constant dc current at 50 A and u_{out} as expected a square wave form with an amplitude of 49.54 A. Therefore, the leakage current in this case is approximately 500 mA. Additionally, the output current $i_{out,sim}$ of the simulation model described in section 4.4.1 with an inductance L_{sc} set to zero is displayed. Compared to the measurement results, the simulation shows the ideal rectangular current curve without commutation time. It can also be seen that the period duration of the measurement, as explained above, is slightly longer than the specified 0.1s. The input and output voltage is shown in fig. 4.17 (b). For both, an induced voltage occurs with each switching operation, which also decreases as the rate of current change reduces. The induced voltage is always positive for the input voltage with an amplitude of around 13 mV. With the output voltage, the induced voltage is opposite to the change in current, i.e. approximately −7 mV when the output current is changing from −50 A to 50 A, and 7 mV for 50 A to −50 A. In steady state, the voltage is zero for both. This behavior is to be expected as the load is a superconductor and therefore has very low impedance. Therefore, the shape of the voltage is determined by the self-inductance of the circuit and the load.

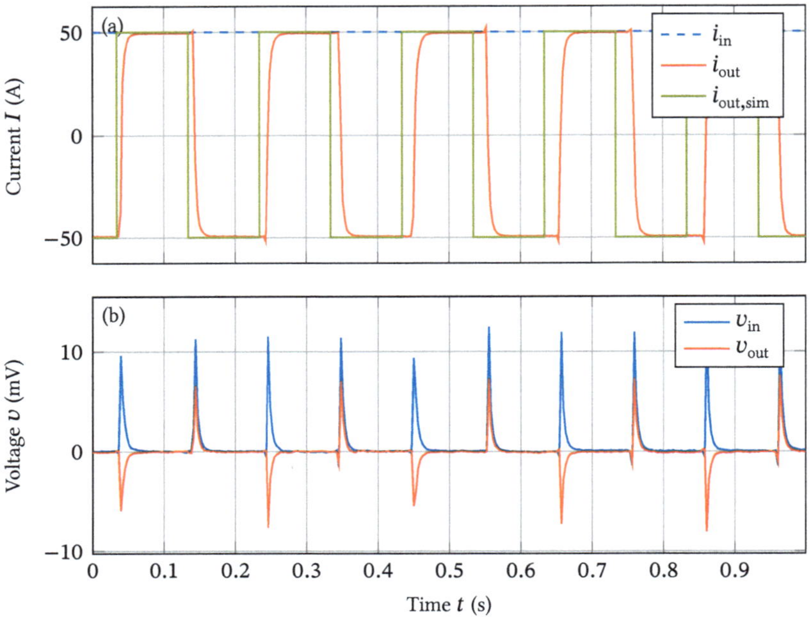

Figure 4.17: Input and output voltage and current in the superconducting bridge inverter during continuous stress test. The electromagnet set point is 1000 Hz, 185 mT, and $t_{sw} = 0.1$ s. Displayed are the input and output currents (a) and the according voltages (b).

4.4 Simulation Model and Scaling Laws

A simplified simulation model is now being developed to show the relationships between leakage currents, load currents and the impedances that occur in the circuit. We now consider the circuit diagram of the bridge circuit in figs. 4.1 and 4.2 and first find an appropriate model for each component to reproduce the behavior.

4.4.1 Simulation Model

Superconducting switches based on the dynamic resistance have two states, either closed or open. When closed, the resistance can be assumed as zero when the transport current is below the critical current which is the case here. When open, the switch takes on a resistance whose value depends on the magnetic field amplitude and frequency, see section 3.3. The switch can therefore be modeled as a resistor that can assume two values depending on the switching state. Any transient behaviors, which would add an inductance for the finite current slew rate, are neglected. It is also assumed that all four switches have the same behavior and therefore identical values, namely a resistance of $0\,\Omega$ and in the open state the value R_{off}. Measured values of the total resistance in section 3.3 reach from $0\,\text{m}\Omega\,\text{m}^{-1}$ to $123.5\,\text{m}\Omega\,\text{m}^{-1}$.

As input a power source is utilized, in this case an ideal constant current source to set a fixed input current i_{in}. The load represents a sink for electrical power. Since no transient processes are considered here either, this is modeled as a simplified resistor with a value of R_{load}. Moreover, voltage drop across the load is determined by the load current and the load resistance

$$v_{\text{out}} = R_{\text{load}} \cdot i_{\text{out}}. \tag{4.1}$$

The general equivalent circuit is displayed in section 4.4.1, where $R1$ to $R4$ are the respective equivalent resistances of the switches $S1$ to $S4$. The load resistance R_{load} is calculated with the output voltage v_{out} and the output current i_{out},

$$R_{\text{load}} = \frac{v_{\text{out}}}{i_{\text{out}}}. \tag{4.2}$$

Furthermore, the resistances $R1$ to $R4$ have a variable value that is dependent on the switching state of the switches. When the switch is in on-state, the superconducting tape is superconducting and therefore the dc resistance is equal to zero. In off-state of the switch, an external alternating magnetic field with the amplitude of B_{a} and the frequency f is applied to the HTS tape. This generates the dynamic resistance R_{dyn} in the superconducting layer and results in the total resistance R_{tot} in which also the resistances of the normal conducting layers are included. This property is tape specific and the according resistance values of the tape investigated in this work can be found in section 3.3. Here the off-state resistance is called R_{off} and is assumed equal for the four switches.

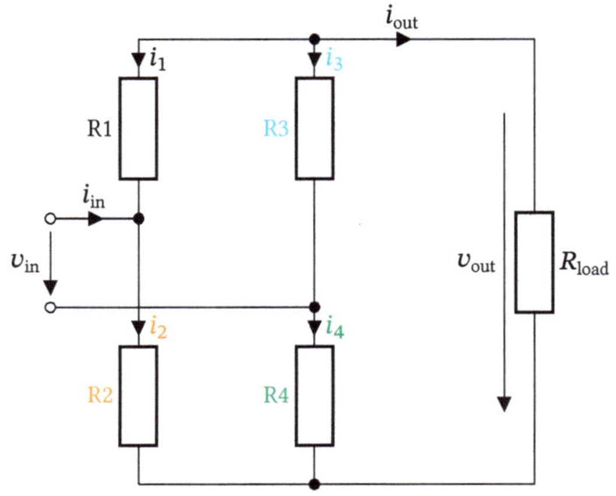

Figure 4.18: Simplified resistive simulation model of the fully superconducting bridge inverter, neglecting all transient behaviors.

This equivalent circuit is then modeled in Simulation Program with Integrated Circuit Emphasis (SPICE) to show the general function of the circuit and displayed in section 4.4.1. Hereby, the input is set to a constant input current source and the output modeled by a constant resistance with the value R_{load}. The switch resistances are time-dependent resistances alternating between R_{on} and R_{off} in a square waveform, where $R1$ to $R4$ have the same waveform. But have a phase shift of 180° for $R2$ and $R3$ similar to fig. 4.15 (top). This is implemented with conditional statements and an additional helper voltage source vh with the specific switch frequency $f_{sw} = 5\,\text{Hz}$ and an arbitrary sinusoidal waveform. For the resistances $R1$ and $R4$, the on-state resistance is defined via R_{on} when the voltage vh is positive and R_{off} when negative, vice versa for the resistances $R2$ and $R3$. However, in the SPICE model R_{on} could not be set to zero due to convergence issues, therefore it is set to 1 pΩ. By adding a self-inductance L_{sc} to all the components, the real commutation behavior with a finite slew rate can be further investigated. In this case, L_{sc} is set to a constant value of 500 nH.

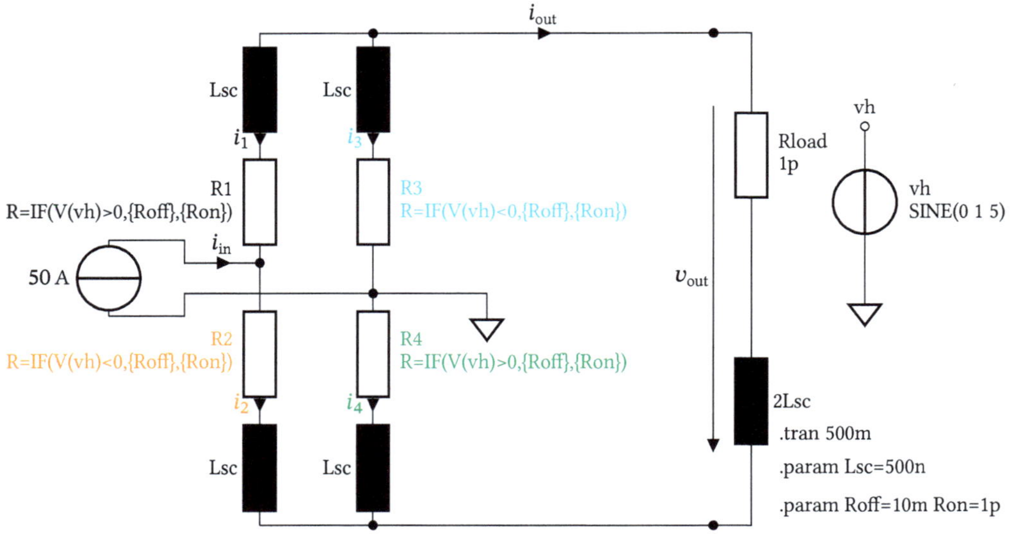

Figure 4.19: Simulation model of the bridge inverter in SPICE including the simulation parameters.

Figure 4.20 shows the voltages and currents of the simulation for an off-state resistance of 100 mΩ. The input and output currents are shown in the top and the voltages in the bottom graph. Both show the expected behavior of a fully superconducting H-bridge inverter. Comparing to the measurement results from fig. 4.17, the simulation has the same qualitative characteristics. As displayed, the dc input current is set to 50 A while the output current has a rectangular waveform alternating between the positive and the negative amplitude of 48 A.

With the voltages, a peak occurs each time the current is switched. In steady state, however, the voltage is not zero as in the measurement due to the small on-state resistance that must be applied to the superconductor for reasons of convergence, which also leads to a higher leakage current. This will be discussed in more detail in the next section.

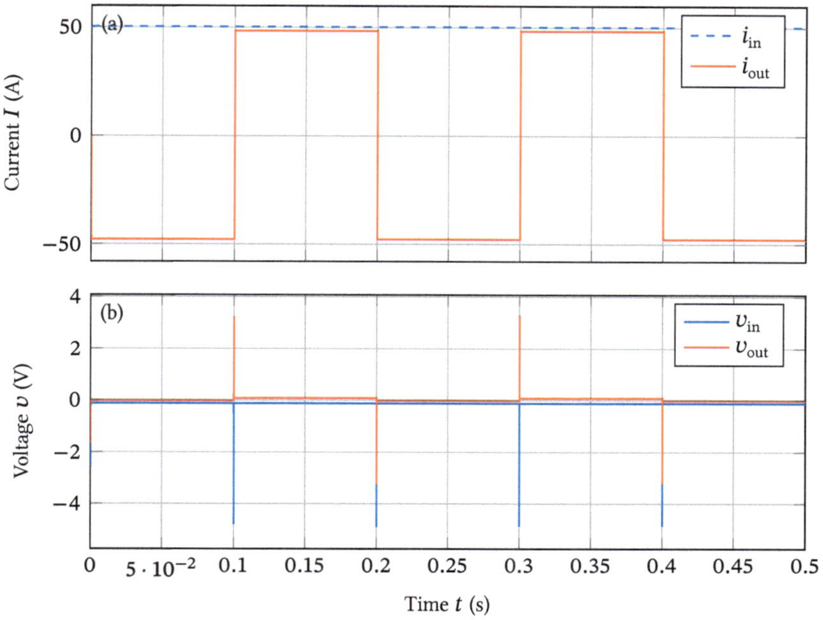

Figure 4.20: Simulated input and output voltage and current of the superconducting inverter according to section 4.4.1 with $R_{off} = 100\,m\Omega$.

In fig. 4.21 the current in switch S1 i_1 is plotted while varying the ratio between off-state resistance and load resistance $\frac{R_{off}}{R_{load}}$ by changing the off-state resistance R_{off}. As clearly visible, the leakage current and therefore also the output current is highly dependent on the ratio between these resistances. Only when R_{off} is high enough compared to the load resistance, the amount of leak current flowing to the switches in off-state is small. With smaller ratios the leak current is getting higher, effectively reducing the output current. Also with higher R_{off} the commutation takes place faster. This simulation is comparable with the measurement in fig. 4.13 where the off-state resistance is varied by modifying the normal conducting layers.

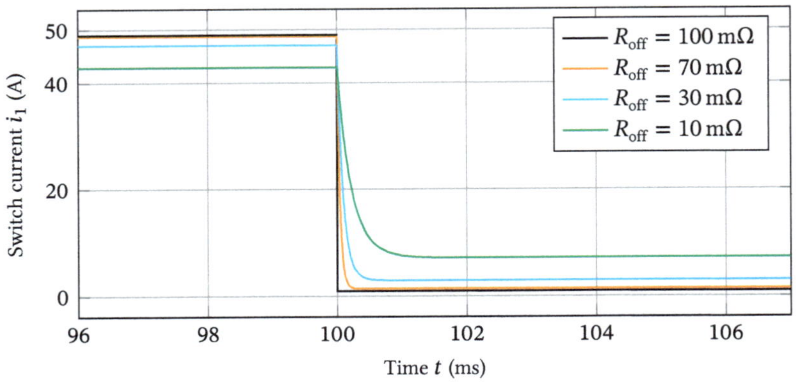

Figure 4.21: Comparison of the commutation behavior of switch S1 in the bridge inverter under variation of the off-state resistance in the SPICE model.

4.4.2 Efficiency Analysis

In stationary operation only states 2 and 3 of the circuit according to fig. 4.2 are relevant. These cases can be simplified with the above findings as pictured in fig. 4.22a and fig. 4.22b. Switches in on-state can be replaced with short circuits and the switches in off-state with the off-state resistance. With the load resistance, this results in three parallel resistances connected to the current source. Since there is no voltage drop across the superconducting switch in on-state, the output voltage u_{out} is equal to the input voltage i_{in}. As it is also assumed that all switches have the same resistance value when switched off, $R1_{\text{off}} = R2_{\text{off}} = R3_{\text{off}} = R4_{\text{off}} = R_{\text{off}}$, the switch off-state resistances can be combined, and the individual leakage currents added to the total leakage current, $i_{\text{leak}} = i_2 + i_3$ for fig. 4.22a and $i_{\text{leak}} = i_1 + i_4$ for fig. 4.22b. This simplified circuit is displayed in fig. 4.22c where the off-state resistance has the value $\frac{R_{\text{off}}}{2}$. The input current i_{in} is divided in

$$i_{\text{in}} = i_{\text{out}} + i_{\text{leak}}. \tag{4.3}$$

The ratio between load current and leakage current is therefore determined solely by the ratio of the resistances

$$\frac{i_{\text{leak}}}{i_{\text{out}}} = \frac{2R_{\text{load}}}{R_{\text{off}}}. \tag{4.4}$$

This means that if the maximum leakage current i_{leak} target is 0.1 % of the load current i_{out}, the off-state resistance R_{off} of the switches has to be higher than $2000 \cdot R_{\text{load}}$ and a leakage current of 1 % means a resistance of $200 \cdot R_{\text{load}}$.

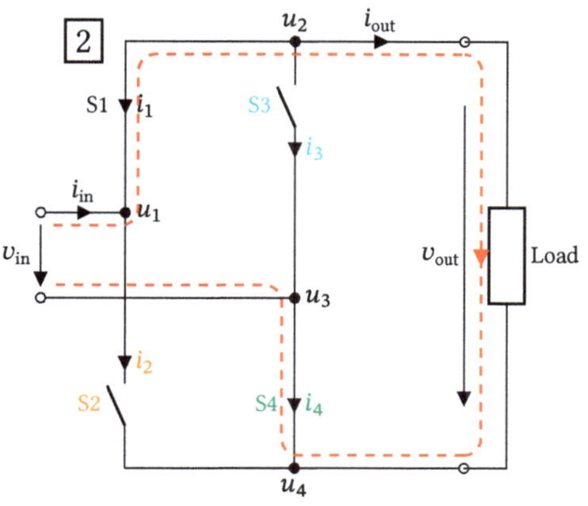

(a) S1 and S4 closed

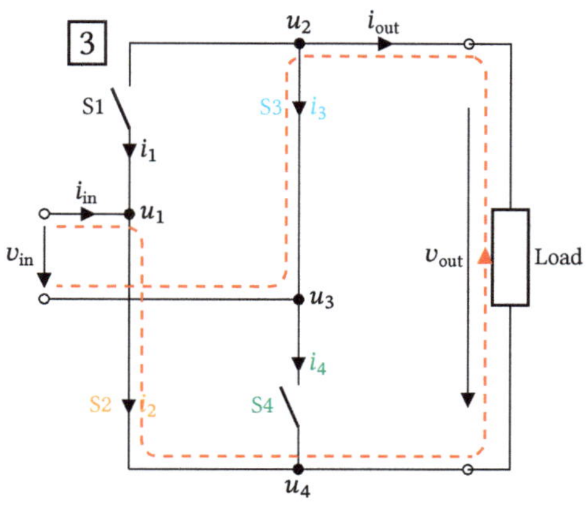

(b) S2 and S3 closed

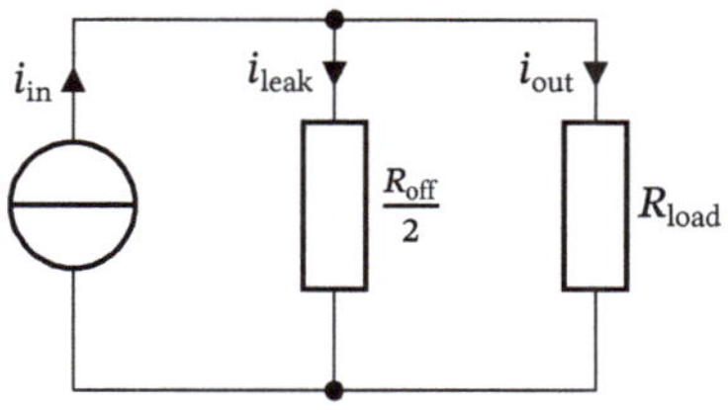

(c) simplified

Figure 4.22: Relevant switching states of the H-bridge circuit to achieve an alternating output current.

These relationships can also be used to derive the electrical efficiency η_{el}. Generally, the electric efficiency is the ratio between output power P_{out} and the input power P_{in}

$$\eta_{\text{el}} = \frac{P_{\text{out}}}{P_{\text{in}}} = \frac{v_{\text{out}} \cdot i_{\text{out}}}{v_{\text{in}} \cdot i_{\text{in}}}. \tag{4.5}$$

Since the resistance in on-state is zero and therefore no voltage drop occurs across the switch in on-state, the output voltage is equal to the input voltage. This simplifies eq. (4.5) with eqs. (4.3) and (4.4) to the ratio between output current and input current

$$\eta_{\text{el}} = \frac{v_{\text{out}} \cdot i_{\text{out}}}{v_{\text{in}} \cdot i_{\text{in}}} = \frac{i_{\text{out}}}{i_{\text{in}}} = \frac{1}{1 + 2\frac{R_{\text{load}}}{R_{\text{off}}}}. \tag{4.6}$$

Equation (4.6) is a proper rational function which is plotted in fig. 4.23 and represents the efficiency as a function of the resistance ratio $\frac{R_{\text{load}}}{R_{\text{off}}}$. The efficiency goes towards 1 for smaller ratios and to zero for larger ratios. At a ratio of exactly 1, the efficiency is 1/3. An efficiency of 0.99, on the other hand, is achieved at a ratio of around $5 \cdot 10^{-3}$. This corresponds to an off-state resistance of $\approx 198 \cdot R_{\text{load}}$, which means that R_{off} must be significantly greater than R_{load} in order to achieve a high efficiency. However, it must be noted that this consideration of the efficiency does not include dynamic losses such as the redistribution of the current as switching losses, and therefore only serves as an initial point of reference. This consideration should serve as a rough estimate of the functionality and to determine the quantity of superconductors for the design of a superconducting inverter.

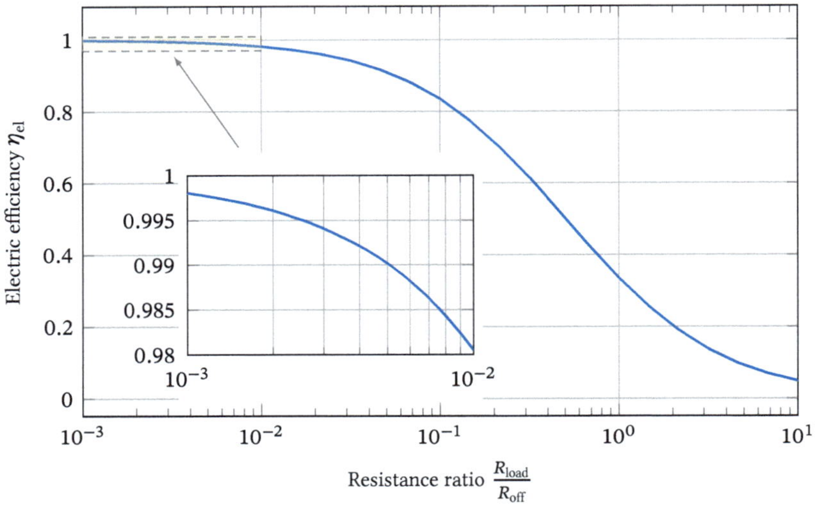

Figure 4.23: Electric efficiency of the current inverter dependent on the ratio of off-state resistance and load resistance.

4.4.3 Scaling Laws

If the inverter is to be used for a specific application, the inverter must be adapted to the required specifications. The output voltage and the output current are usually specified, and therefore also the output power. In addition, the configuration depends on the properties of the superconductor used.

In on-state of a switch it is important, that the superconductor can carry the load current without exceeding its working parameters. This can be either done by utilizing wider HTS tapes with a higher critical current or connecting multiple tapes parallel to reach the required current capacity. The number of parallel superconductors per switch is determined by the critical current of each tape and the design current capacity. However, it is crucial to add a safety margin to allow over-current situations without damaging the HTS tapes, e.g. setting the working current to 80 % of the tape's critical current. The number of parallel tapes N_{pt} can be calculated by

$$N_{pt} = \frac{i_{out}}{\gamma I_c} \tag{4.7}$$

where i_{out} is the design output current amplitude of the inverter, I_c the critical current of each individual tape, and γ the safety factor which determines the operating point of each parallel tape. In addition, other external influences that can affect the critical current when switched on must be taken into account, for example the close physical proximity to a ferromagnetic material, as used for the electromagnets in this work. This ultimately results in the number of parallel superconductors in one switch.

In order to be able to build up a voltage across the load, the switches are considered in the off-state. These must build up a certain resistance value in relation to the output load to enable an effective power flow while constraining the leakage current i_{leak} via the switches in off-state, see fig. 4.23. The minimum resistance value $R_{off,min}$ that the switches must have when switched off so that the specified maximum leakage current i_{leak} is not exceeded, can be calculated using eq. (4.4) with R_{load} defined as

$$R_{load} = \frac{v_{out}}{i_{out}}. \tag{4.8}$$

However, since several superconductors are now connected in parallel to achieve the required current carrying capacity, the total resistance of the parallel circuit $R_{p,tot}$ must reach the minimum off-state resistance and is calculated by

$$\frac{1}{R_{p,tot}} = \frac{1}{R_{off,1}} + \frac{1}{R_{off,2}} + \cdots + \frac{1}{R_{off,npt}} \tag{4.9}$$

with N_{pt} parallel connected HTS tapes. Assuming identical tapes and therefore the same resistance $R_{off,npt}$ for all parallel tapes the equations simplifies to

$$R_{off,min} = R_{p,tot} = \frac{R_{off,npt}}{N_{pt}} \tag{4.10}$$

where $R_{off,npt}$ is the resistance which each parallel connected tape has to build up in off-state of the switch to achieve the minimal required off-state resistance $R_{off,min}$.

The resistance in the superconductor is based on the dynamic resistance, see section 3.3, and is therefore unique for each superconductor model and dependent on the applied external magnetic field amplitude and frequency, which need to be obtained beforehand. Based on these results, the total length of superconductor in active magnetic field, which is needed to build up the minimal required resistance, can be obtained and multiplied by the number of parallel strands. This means that the efficiency also influences the required superconductor length due to the resulting maximum leakage current i_{leak}. The higher the efficiency η_{el}, the lower is the leakage current and the higher is the off-state resistance $R_{off,min}$ and, accordingly, the longer the HTS tape that is in the magnetic field. The self-inductance of the inverter is determined by the shape of the HTS tapes. In the case of a design with straight superconducting tapes, the inductance can be approximated by eqs. (3.6) to (3.8).

To summarize, first the current carrying requirement is met by connecting the appropriate number of HTS in parallel, then the voltage and efficiency constraint is fulfilled by building up a suitable off-state resistance which is dependent on the tape's total resistance, length, and number of parallel HTS tapes.

4.5 10 V/10 kA–Demonstrator Design

Using an example application, the design of a fully superconducting inverter is now demonstrated. This is an application with low voltages and high currents in the 0.1 MVA range, as could be used for the electrolysis of water, for example, where a cell voltage of around 1.5 V is required [Shi22]. The superconducting inverter is now to be designed so that the output voltage is 10 V at a load current of 10 kA. As before, a purely ohmic load is assumed, whose resistance R_{load} is 1 mΩ. With the target of a maximum leakage current of 10 A, i.e. 0.1 % of the output current, the off-state resistance of the utilized superconducting switches has to be 2 Ω according to eq. (4.4). Consequently, the efficiency is 99.9 % with eq. (4.6). When setting the efficiency to 99 %, the leakage current is 101 A, which results in a minimum off-state resistance $R_{\text{off,min}}$ of approximately 0.2 Ω. The parameters are summarized in section 4.5.

Table 4.3: Key requirements for the 10 V/10 kA-demonstrator

Symbol	Parameter	Value
v_{out}	output voltage	10 V
i_{out}	output current	10 kA
η_{el}	el. efficiency	99 % and 99.9 %
i_{leak}	max. leak current	10 A
B_{a}	magnetic field amplitude	207 mT
f	magnet frequency	1000 Hz

4.5.1 Demonstrator Design with the Investigated HTS Tape

The requirements for the four superconducting switches can be derived from these parameters. In on-state, each switch needs a current carrying capacity of 10 kA, while in off-state a resistance of at least 1.998 Ω is required for 99.9 % efficiency. To simplify matters, a resistance of 2 Ω is assumed. Accordingly, for 99 % efficiency the resistance is 0.2 Ω.

Step 1: Meeting the current specifications

The first condition determines the number of superconductors connected in parallel in the switch. If the superconducting tape examined here in this work is taken, it has a minimum critical current of 338 A, see table 3.2. Defining the operation point of the superconductor at 80 % of its critical current and therefore having a safety margin of 20 % means a current carrying capacity of 270.4 A. This leads to 37 parallel connected HTS tapes for each switch to reach 10 kA. This applies to both examined efficiencies. It is to be noted that no further reduction of critical current by the arrangement of the HTS tape is considered [Byk14].

Step 2: Meeting the voltage specifications

The specified output voltage and the maximum leakage current determine the level of resistance of the switch in off-state. To meet the requirements above, the minimal off-state resistance $R_{\mathrm{off,min}}$ of each switch is 2 Ω for 99.9 % efficiency and 0.2 Ω for 99 % according to eq. (4.4). To reach this resistance with 37 HTS tapes connected in parallel, each individual superconductor requires an off-state resistance $R_{\mathrm{off,npt}}$ of 74 Ω which is built up by the dynamic resistance effect. Accordingly, in case of 99 %, $R_{\mathrm{off,min}}$ is 0.2 Ω, and therefore each individual tape needs a resistance of $R_{\mathrm{off,npt}} = 7.4\,\Omega$.

Step 3: Calculating tape length for one switch

Considering the measured total resistance of 119.8 mΩ m^{-1} at 207 mT and 1000 Hz for configuration C in section 3.3 from fig. 3.17b, the length of each superconducting tape is calculated to 617.70 m for $\eta_{\mathrm{el}} = 99.9\,\%$ to reach the required 74 Ω. Summing up the parallel tapes of one switch would mean a total length of 22.85 km in active magnetic field. In case of $\eta_{\mathrm{el}} = 99\,\%$, the lengths are 10 times shorter due to the ten times lower $R_{\mathrm{off,npt}}$, so the total length per switch is 2.26 km. As can clearly be seen, the current carrying capacity and therefore the number of parallel tapes has a major influence on the required tape length.

Step 4: Calculating tape length for the whole circuit

Since four switches are needed for the inverter, the total amount of superconductors in the magnetic field is 91.42 km for 99.9 % efficiency. Additional tape length is needed for the interconnection of the four switches, which is assumed as 10 % of the active magnetic field length. In total, this results in approximately 100.56 km of tape investigated in this work. Assuming a price for superconductors of 50 €/m, the amount needed for the inverter would cost 5.02 M€.

For 99 % efficiency, the total tape length is summed up to 10.06 km, one tenth compared to $\eta_{el} = 99.9\,\%$. Likewise, the costs, which are estimated at 500 k€.

4.5.2 Impact of Wider Tapes

As the demonstrator design above shows, a large quantity of superconducting tapes is needed for a 100 kVA-class inverter with nowadays readily available HTS tapes. One of the main reasons for this is the high number of HTS tapes connected in parallel to achieve the required current. The number of parallel tapes has an influence on the resistance and therefore also on the length of each individual HTS tape. Hence, a possible optimization could be to increase the current carrying capacity of an HTS tape, for example with wider tapes. Prototypes and research data are already available from different manufacturers with the prospect of 40 mm to 120 mm wide HTS tape, i.e. ten times the width investigated in other works [Uso17, Har20, Lee23, Hol25]. In the following, a 48 mm wide HTS tape with the same properties is assumed, i.e. four times the width and the same design steps as above are conducted for an efficiency of 99.9 %. The results for 99 % efficiency are listed in table 4.6.

Furthermore, it is assumed that the critical current per width remains the same and therefore the 48 mm HTS tape has a critical current that is four times higher than the 12 mm tape of 1352 A. When the operating point is targeted at 80 % of the critical current, only 10 parallel tapes are required to reach a current capacity of 10 kA. The lower number of parallel tapes also reduces the required resistance of each tape $R_{off,npt}$ to 20 Ω in order to reach 2 Ω off-state resistance of the switch. A simultaneous increase in the width and increase in the critical current by the same factor does not lead to any changes in the dynamic resistance R_{dyn} according to eq. (2.13). If we now assume a substrate with a thickness of 25 µm, i.e. 25 % of the initial thickness, the normal conductive resistance R_{nc} remains the same when the width is increased and thus also the total resistance R_{tot} of $119.8\,\mathrm{m\Omega\,m^{-1}}$ in off-state. To reach the minimal required off-state resistance $R_{off,min}$ of the switch, each individual parallel tape needs a length of 166.94 m in active magnetic field, in total 1.67 km for each switch. Summing up the required amount of HTS tape for all four switches plus the 10 % of interconnecting tape, this results in 7.35 km of 48 mm wide tape.

To conclude, taking a four times wider tape and therefore having a four times higher critical current, reduces the amount of superconductor needed from 100.56 km of 12 mm wide tape to 7.35 km of 48 mm wide tape. This means a reduction of approximately 92 % of tape length. As no price information can be found for wider HTS tapes, a price four times higher is assumed here, this means 200 €/m. Accordingly, the costs in this case amount to 1.47 M€, which is only about 30 % of the costs of the inverter with a 12 mm tape.

4.5.3 Impact of Thin Films on Sapphire Substrate

Another interesting option are superconductive ReBCO thin films deposited on a sapphire substrate [Kra03, Deu18, Gaw19a]. The metallic substrate that is normally used is substituted with a ceramic aluminum oxide (Al_2O_2) wafer, also called sapphire. The other layers remain similar. There is a buffer layer between the ReBCO layer and the substrate and a top cap layer is applied to the superconducting layer; in the case of a thin film on sapphire, this often consists of gold (Au). The sapphire substrate offers a higher stability with its higher thermal conductivity and heat capacity compared to conventional substrates. Sapphire as substrate has already been investigated for use in fault current limiters [Gro99, Deu18, Pop22, Tix23]. In addition, sapphire has a resistivity of about $1 \cdot 10^{16}$ Ω cm at RT and thus can be neglected in the normal conducting resistance. Figure 4.24 displays an HTS thin film on sapphire substrate and with a Au cap layer on top manufactured by ceraco [cera]. Comparing silver and gold in figs. B.2 and B.3, gold has about 1.7 times higher resistivity at 77 K than silver. This would also contribute to a higher resistance. The dimensions are given in the schematic drawing below and the layer thicknesses and several key parameters are listed in table 4.4.

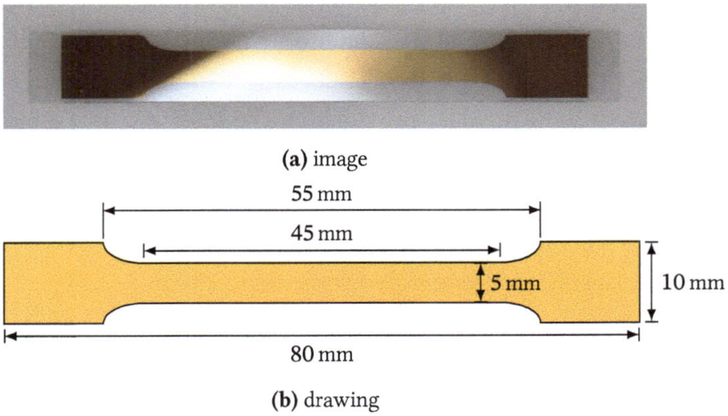

Figure 4.24: Image and drawing of the thin film sample on sapphire substrate for the critical current measurement. The schematic below shows the pattern of the superconductive thin film layer on the rectangular shaped substrate.

Table 4.4: Parameters of the thin film superconductor on sapphire substrate manufactured by ceraco

Symbol	Parameter	Value
	model	M-type
	substrate material	Al_2O_3 wafer
h_{rebco}	ReBCO film thickness	330 nm
h_{subs}	substrate film thickness	50 µm
$h_{Au,top}$	Au top film thickness	200 nm
T_c	critical temperature	87.1 K
j_c	critical current density	3.3 MA cm^{-2}
I_c	measured critical current	61 A

With the width of 5 mm and the only normally conductive gold layer with a thickness of 200 nm, the result is a normal conducting resistance R_{nc} of 4.552 Ω m^{-1} at 77 K from fig. B.3 which is approximately 100 times higher than for the HTS tape investigated in this work in config. C, see section 3.2.3. However, the low thickness of the ReBCO layer of 330 nm results in a lower critical current compared to conventional HTS tapes. Figure 4.25 shows the critical current measurement of the thin film sample displayed in fig. 4.24. For that, flexible copper current leads are soldered to both 10 mm wide ends of the sample with InSn solder. The voltage tapes are clamped at a distance of 2.5 cm in the center of the sample. The I–E–curve is then fitted to eq. (3.5) with the results $I_c = 61.00$ A, $n = 45.20$, $c_1 = 1.77 \cdot 10^{-8}$, and $c_0 = -2.95 \cdot 10^{-7}$. The results correspond well with the data provided by the manufacturer giving a slightly higher critical current density of 3.70 MA cm^{-2} than listed. In Gawith et al. [Gaw19a] the dynamic resistance of thin films from the same manufacturer has been investigated on a 10 mm wide sample showing high built-up resistances of 300 mΩ m^{-1} at 50 mT and 8.8 kHz, which is higher than the theoretical 166 mΩ m^{-1} calculated from eq. (2.13).

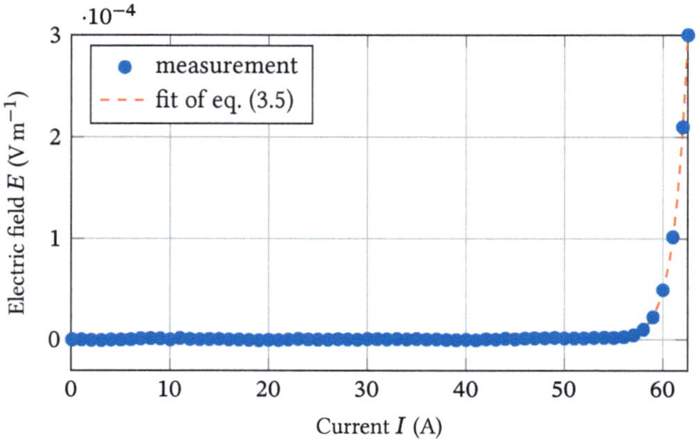

Figure 4.25: Critical current measurement of the ReBCO thin film samples on sapphire substrate in fig. 4.24.

With the data from above, the demonstrator design, based on superconducting thin films on sapphire substrate, is conducted for the widths 10 mm and 40 mm with the critical currents 100 A and 400 A respectively and an off-state resistance of 300 mΩ m^{-1}.

To reach the design current of 10 kA with 10 mm wide thin film, 125 parallel superconductors are needed with the current capacity of $0.8I_c$. Therefore, each parallel conductor needs a resistance of 250 Ω to obtain the required minimum off-state resistance $R_{\text{off,min}}$ of 2 Ω. This can be achieved with 125 times the section of 833 m of thin film in magnetic field, which sums up to 104.17 km per switch. Multiplying by four and adding the 10 % for connecting the tapes totals to 458.15 km. Assuming a price of 500 €/m for 10 mm wide thin film [Tix23], the cost of the superconductors would be 229.2 M€.

The same is done for 40 mm wide thin film, where current density, cost per width, and dynamic resistance is the same as for the 10 mm wide thin film. Since the current capacity is higher, only 32 conductors are needed in parallel, adding up to 6.83 km per switch and 30.04 km in total. This means a reduction of thin film length of approximately 93 % only by increasing the width by the factor of 4. The cost is also reduced to 60 M€.

The results of all design variants are summarized in table 4.5. The comparison shows that despite the advantageous higher off-state resistance of the sapphire based thin films, more superconductor lengths are required. This is due to the lower YBCO layer thickness and the associated lower critical current, which leads to more conductors connected in parallel. Wider tapes are also preferable, as these also reduce the number of parallel conductors. As for now, the HTS tapes are superior to the sapphire thin films in terms of overall length and cost. However, deposition on sapphire offers the potential to build up higher voltages than the tapes due to the greater off-state resistances. Disadvantageous are the low ReBCO layer thicknesses, which are limited due to the different thermal expansion coefficients of ReBCO and sapphire. Increasing the deposited layer thickness would lead to thermal cracking. Other substrate materials such as lanthanum alumina ($LaAlO_3$), yttria-stabilized zirconia (YSZ), or magnesium oxide (MgO) have similar electrical properties to sapphire but allow greater layer thicknesses of up to 3 µm. However, these are considerably more expensive than sapphire. [Phi96, Sem02, Guo13, Zha19b, San23, The25, cerb]

Table 4.5: Summary of 10 V/10 kA demonstrator with different superconductor choices with 99.9 % efficiency.

Parameter	**SF12100**	**SF12100**	**Sapphire**	**Sapphire**
width (mm)	12	48	10	40
critical current (A)	338	1352	100	400
off-state resistance ($m\Omega\, m^{-1}$)	119.8	119.8	300	300
number of parallel tapes per switch	37	10	125	32
sc length per switch (km)	22.85	1.67	104.17	6.83
total length (km)	100.56	7.35	458.33	30.04
cost (M€)	5.02	1.47	229.2	60.08

The same calculations have been conducted for an efficiency of 99 %. According to eq. (4.6), the minimum off-state resistance $R_{\text{off,min}}$ in this case is 0.198 Ω, thus 0.2 Ω for simplicity is chosen. The results are given in table 4.6. Compared to the results of 99.9 %, the HTS tape lengths are consistently ten times smaller, as the leakage current is also assumed to be ten times greater. This is also reflected in the ten times lower costs, with the conventional 48 mm HTS as the most cost-effective variant. It should be noted that there are currently already 12 mm wide superconducting tapes that reach a critical current of over 800 A at 77 K and self field [Mol25, Son25]. These would therefore require far fewer parallel conductors to carry the current and would therefore also lead to different conductor lengths and costs.

Table 4.6: Summary of 10 V/10 kA demonstrator with different superconductor choices with 99 % efficiency.

Parameter	**SF12100**	**SF12100**	**Sapphire**	**Sapphire**
width (mm)	12	48	10	40
critical current (A)	338	1352	100	400
off-state resistance ($m\Omega\, m^{-1}$)	119.8	119.8	300	300
number of parallel tapes per switch	37	10	125	32
sc length per switch (km)	2.26	0.167	10.42	0.683
total length (km)	10.06	0.735	45.83	3.00
cost (M€)	0.50	0.15	22.92	6.01

4.5.4 Generating Long Length Magnetic Fields

As previously shown in this chapter, the various demonstrator designs require magnets that generate a time-varying magnetic field over long lengths. This is not possible with the magnets from section 3.1.1, as they only generate the magnetic field in the air gap over a length of 2 cm. In order to have longer lengths of superconductors in the magnetic field, there are already various approaches in literature [Jia17b, Jia18, Li19] which incorporate winding or stacking of HTS tapes. One example is the concentric iron core in [Li18], in which the superconducting tape is wound around a round iron core in bifilar manner and an air gap is formed with the help of an outer iron core. The magnetic field is generated via a copper winding around the inner iron core. In this case, 3.3 m of 6 mm wide tape was loaded with an alternating magnetic field. Another method to generate the magnetic field would include a superconducting coil without an iron core, such as the rectangular disk-up-down assembly (rDUDA) coil [Arn21, Wan23, dOli24]. These are constructed similarly to Bitter electromagnets [Zhu21] but in a rectangular shape with HTS tapes by stacking and connecting the tapes.

Figure 4.26 shows the exploded view of such winding with four HTS tapes. The tapes are connected via soldered face-to-face contacts at one end, providing a circular current path around the c-axis of the tapes. The inner bore of the windings is laser cut into the tape and can be varied according to the needs. This technology enables the generation of magnetic fields over a long length, since long lengths of HTS tapes are already commercially available.

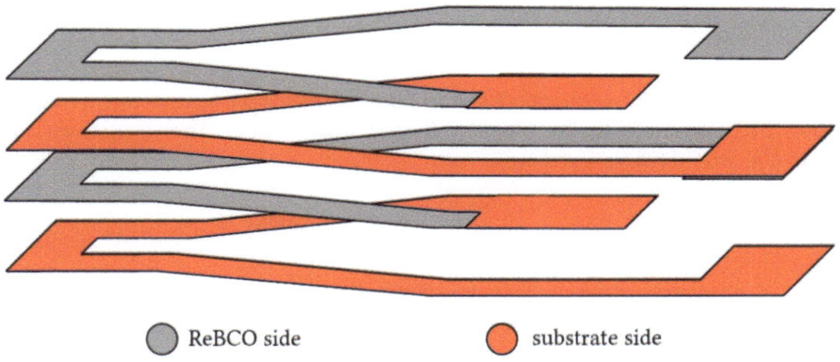

Figure 4.26: Exploded view of the rectangular disk-up-down assembly with the contact area concentrated at the side of the assembly. Grey marks the HTS layer side and orange the substrate side. This arrangement ensures low contact resistances with face-to-face solder connections.

In this work, the generation of a magnetic field over a length of 10 cm with 12 mm wide tapes is investigated, and therefore the HTS tapes are structured according to the drawings in fig. 4.27, where fig. 4.27a shows the pattern for the inner windings and fig. 4.27b a slightly longer variant for the top and bottom tape of the stack for connecting the rectangular disk-up-down assembly to a power supply. The HTS tapes for the inner windings have a length of 200 mm in total with a rectangular shape of the size 100 mm times 4 mm. This reduces the effective width to $\frac{1}{3}$ and thus the critical current. The remaining section of the tape is utilized for soldering the individual windings with SnPb37 solder. The different lengths are selected so that the contact area is approximately the same at around 269 mm^2. The overlap of the individual tapes from the inner soldering joints is over a length of 67.4 mm and a width of 12 mm and the outer soldering joints have an overlap over the entire width of 4 mm and a length of 22.4 mm. The key parameters of the assembly investigated in this work are listed in table 4.7.

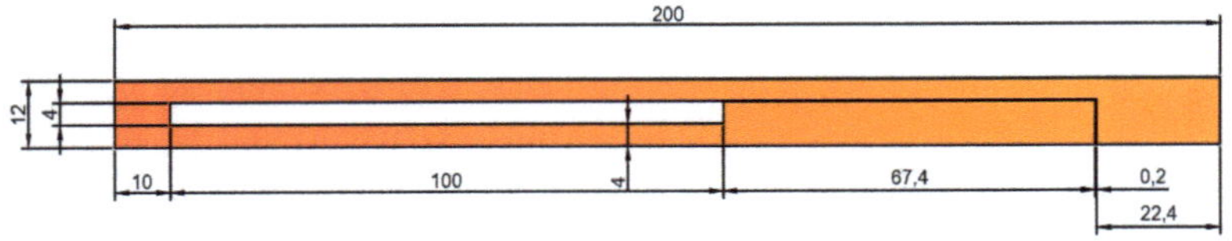

(a) inner winding

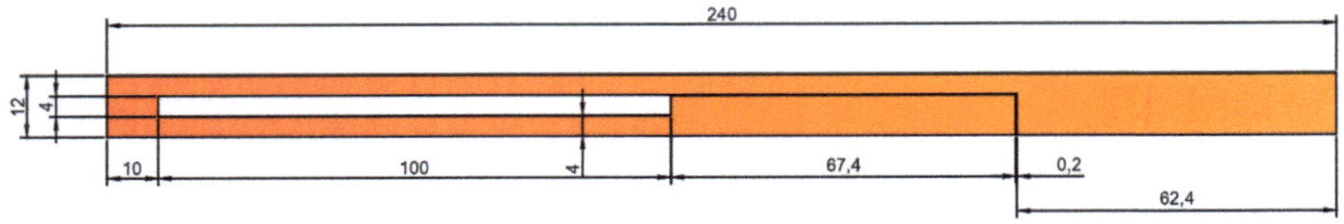

(b) outer winding

Figure 4.27: Schematic drawing of the segments of the rectangular disk-up-down assembly with all dimensions. The top shows the structure of the inner tapes and the bottom shows the outer tapes with an additional section to contact the assembly.

Table 4.7: Key design parameters of the investigated rectangular disk-up-down assembly.

Symbol	Parameter	Value
	SC model	SCS12050
	number of turns	10
	number of joints	9
w_g	gap width	4 mm
l_g	gap length	100 mm
$I_{c,min}$	minimum critical current	100 A
I_{coil}	design coil current	100 A
B_a	design magnetic field at I_{coil}	100 mT

As HTS tape for the rectangular disk-up-down assembly the model SCS12050 manufactured by Superpower is used. This provides a small substrate thickness of 50 µm and additional copper stabilizing. The small thickness of the conductor is advantageous as it means that the tapes in the stack are closer together and a higher magnetic field can be reached. The individual tapes used were also characterized in advance and only those with a critical current of at least 550 A were used, as the current carrying capacity is reduced by two-thirds due to the structuring.

First, the entire stack is soldered using a die and a heating plate. Second, individual tapes are masked with polyimide tape to prevent unwanted soldering and assembled with 50 µm thick solder foil and flux without pre-tinning of the soldering areas. The polyimide tape also acts as turn-to-turn insulation. Third, the entire assembly is soldered under spring pressure and according to the temperature profile of the solder used. The resulting stack with a total of ten windings is displayed in fig. 4.28 with a slight discoloration of the copper due to the heat ingress during soldering.

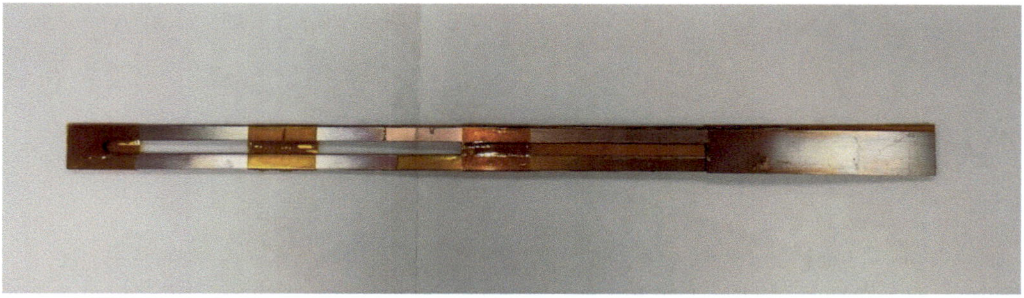

Figure 4.28: Image of the soldered rectangular disk-up-down assembly stack with ten tapes. The inner window of the winding with the size of 100 mm × 4 mm is clearly visible.

Afterward, coil is spread out to mount voltage taps on each winding and measure the critical current of the assembly and contact resistances of the joints. The voltage taps are mounted on the 10 mm wide side opposite to the solder joints, see fig. 4.27a, and the voltage across each joint is measured differential. Additionally, the voltage drop across the current lead is acquired resulting in a total of 10 voltages. Additionally, the test setup is submerged in LN_2 and the current step-wise increased until the nonlinear power law, see eq. (2.1), of one HTS tape is observed to prevent damage to the stack. The current is increased until 103.5 A.

The results are plotted in fig. 4.29. Since the stack consists of 10 HTS tapes, the voltage of 9 solder joints and the total voltage across the current leads are shown. As expected, the voltages show a linear increase, which is due to the resistive solder connection, whereby the gradient, and therefore the joint resistance, differs slightly. Only the voltage v_8 increases in the 80 A to 90 A range to non-linearity and is therefore close to the critical current, while the other voltages continue to increase linearly. Assuming that the critical current is limited within the 26.74 cm long section with a width of 4 mm, this results in a critical current of approximately 98 A. The total voltage v_{tot} is the sum of all voltages. The slope is analyzed in the range of 0 A to 80 A to determine the joint resistance of each tape and the results are listed in table 4.8 with the corresponding specific contact resistances. The majority of the joint resistances are between 120 nΩ to 165 nΩ , two are better and are below 55 nΩ and with joint #9 there is an outlier at 324.8 nΩ. Furthermore, resistance of the rectangular disk-up-down assembly is in total 1.25 μΩ with an average joint resistance of 139.1 nΩ. Moreover, the specific contact resistance of each joint in relation to the area are in a similar range to measured data from literature [Kim13, Léc15], except for the outlier.

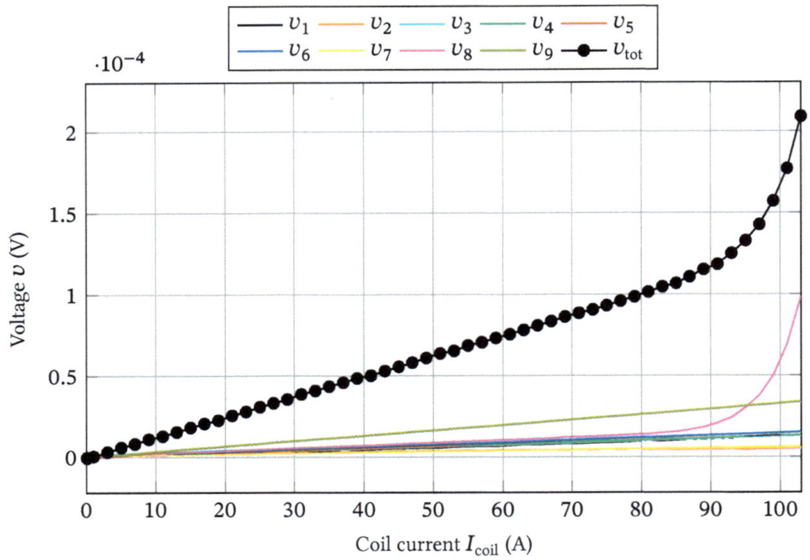

Figure 4.29: Measured voltages across the joints and the total voltage measured near the current leads dependent on the applied current. The voltage v_8 shows a preliminary non-linear increase while the remaining voltages show no component associated with the superconductor.

Table 4.8: Evaluated contact resistances from the voltage-current curves in fig. 4.29. Additionally, the specific contact resistances of each joint are stated.

Joint	#1	#2	#3	#4	#5	#6	#7	#8	#9	Total
resistance (nΩ)	122.4	36.2	134.5	124.9	151.9	141.6	52.3	163.8	324.8	1252.2
specific contact resistance ($\mu\Omega\,mm^2$)	32.9	9.7	36.2	33.6	40.9	38.1	14.1	44.1	87.4	

To verify the built-up magnetic field of the assembly, the applied current I_{coil} is varied to magnetize and demagnetize the coil while measuring the magnetic field B_a of the assembly. For that, the test setup pictured in fig. 4.30 is used where the current leads powering the stack are on the right side. The rectangular disk-up-down assembly is held compact by polyimide tape and the Hall effect sensor, which have been already used in section 3.1.1 to measure the curve of the electromagnets, is fixated in the middle of the 4 mm wide bore of the assembly. Afterward, coil current is increased up to 100 A and then decreased again, resulting in the magnetization curve displayed in fig. 4.31 where the measurement points are pictured as marks (●) and the linear fit as a line (—). The results show a slight hysteresis between magnetizing and demagnetizing while having a linear correlation between current and magnetic field. All in all, the assembly reaches a peak magnetic field of 108.8 mT at a coil current of 100 A. So, the result shows the expected behavior of an air core coil.

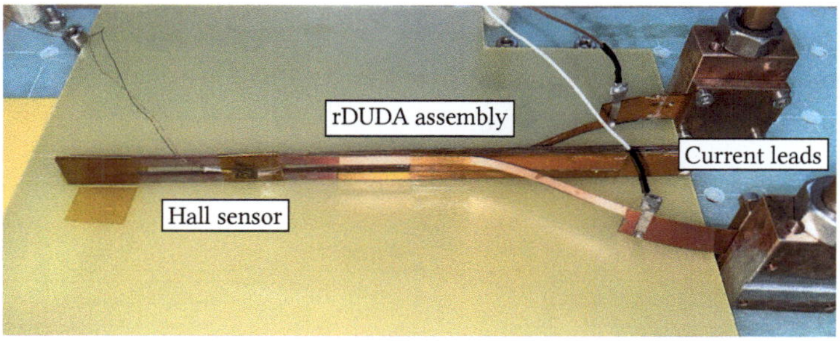

Figure 4.30: Image of the test setup to measure the generated magnetic field amplitude of the rectangular disk-up-down assembly with a Hall sensor.

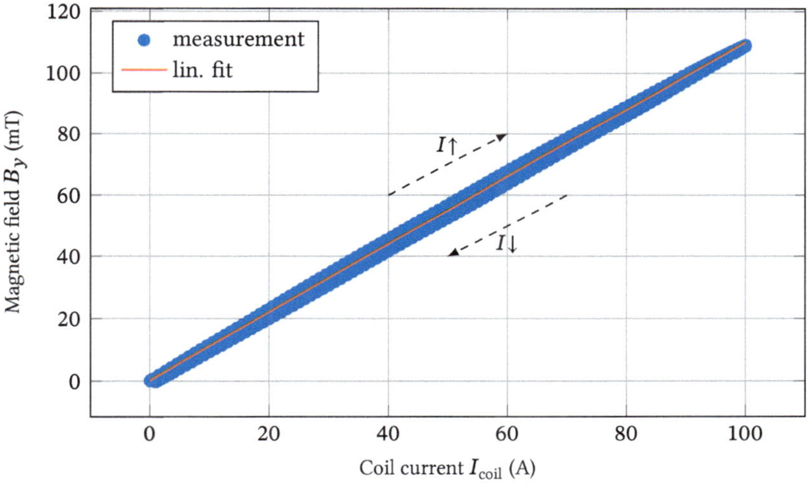

Figure 4.31: Measured magnetic field of the rectangular disk-up-down assembly dependent on the applied current. The hysteresis between magnetization and demagnetization is visible.

To verify the measured values of the rectangular disk-up-down assembly, a 2D FEM model is created in Comsol Multiphysics® to calculate the generated magnetic field. The cross-section is modeled by two stacks of tapes next to each other. In this specific case, each stack consists of ten 4 mm wide tapes placed on top of each other. The distance between the two stacks is 4 mm. For simplification, the thin strip approximation is applied, where the tapes are approximated by current carrying lines which are placed at a distance of 75 µm from each other. The same dc current perpendicular to the cross-section is then applied to each tape of one stack via the surface current density. This is also applied to the second stack, but with a negative sign. The current density is assumed constant along the edge of the HTS tapes. The magnetic field is then analyzed on the 12 mm wide line 50 µm below the assembly, whereby the middle of the line, i.e. at 6 mm, is the approximate position of the Hall sensor. A schematic drawing of the simulation model is displayed in fig. 4.32 with the two 4 mm wide stacks and their current flow direction. Each black line represents an HTS tape and the red line is the location along which the magnetic field amplitude is calculated.

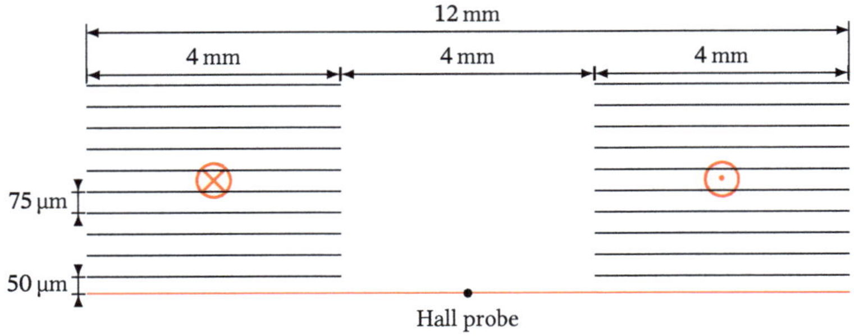

Figure 4.32: Schematic drawing of the rectangular disk-up-down assembly in the 2D FEM simulation. The circles represent the direction of current flow of each stack and the red line displays the position where the magnetic field is analyzed and displayed in fig. 4.33.

The results of the simulation are plotted in fig. 4.33 where the magnetic field norm B_{abs} is plotted along the red line from fig. 4.32 for applied currents between 40 A and 100 A. Moreover, the magnetic field profile is as expected and is highest at the two edges of the bore in the middle and lowest in the middle of the bore. Values of up to 180 mT at 100 A are reached at the edges. The magnetic field flattens out again towards the outer edges of the assembly. It can also be seen that the magnetic field increases linearly with the current. Additionally, the measured values from fig. 4.31 are displayed as marks at $x = 6$ mm. These show good agreement with the simulated results.

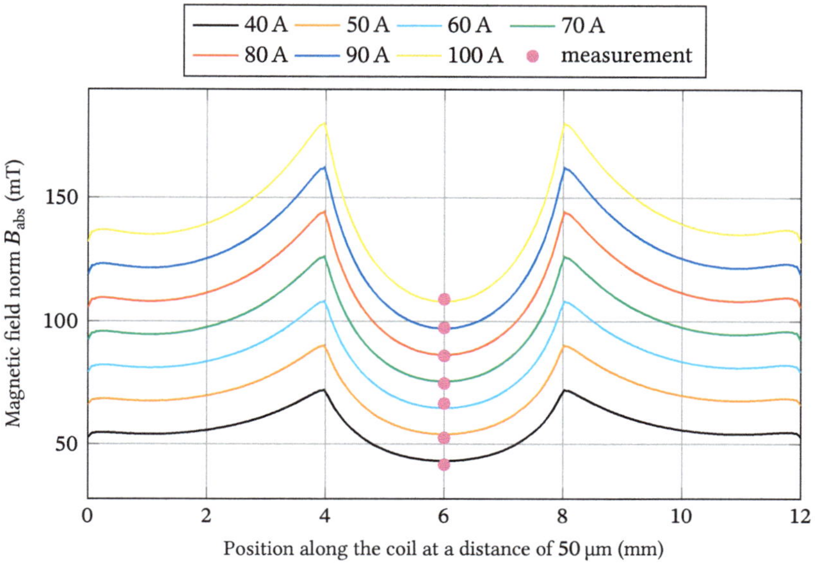

Figure 4.33: Dc magnetic field of the measurement and 2D FEM simulation of the cross-section dependent on the current in the coil. Displayed is the magnetic field norm parallel to the ab-plane at a distance of 50 µm.

These results show an easily scalable method for generating a local magnetic field in the three-digit mT range, which is already sufficient for a long length superconducting switch. The superconductor to be used as a switch would be at the position of the red line in fig. 4.32. If higher fields are required, a sandwich structure with two rectangular disk-up-down assemblies would be possible. For this, the structure in fig. 4.32 is mirrored at the red line with the superconductor between two stacks. This would result in a field that is twice as high with the same applied current. Also, the field peaks at the edges would also decrease, as the x-component of the field would approach zero along the red line. However, these experiments only included dc currents. For building up dynamic resistance an alternating magnetic field is required that needs to be further explored.

4.6 Summary

In this chapter a fully superconducting H–bridge inverter is designed and built by utilizing the results of the preliminary experimental investigations. Initially, the overall function of the inverter is presented, including the three switching states of the superconducting switches.

Section 4.2 presents the HTS tape specific design considerations for the circuit. One major criterion is minimizing contact resistances throughout the entire circuit. This is achieved by using a similar approach to section 3.4 by structuring the HTS tape with a laser to form the required four superconducting switches from a single tape, thereby ensuring superconducting joints between the switches. Initial experiments show an equal distribution of the total critical current of 410 A to both branches of the structured tape with 210 A and 200 A.

In section 4.3 the general functionality of a fully superconducting H-bridge inverter is demonstrated by converting a direct current into a rectangular shaped output current. Additionally, the transient behavior is investigated by switching between the different states. The methods for minimizing commutation times in section 3.2.3 are applied, with their effectiveness demonstrated by achieving commutation times as low as 2.32 ms with variation of magnetic field amplitude and frequency and reducing the leakage current in the switches in off-state to below 100 mA. To demonstrate reliable operation, continuous switching tests were conducted in section 4.3.4, with the current of each switch recorded. The test duration were extended up to 2 min of runtime while varying the pulse width of the trigger signals from 1 s to 25 ms without any change in commutation behavior or instabilities. These results showed that the switches were able to cool down and stabilize sufficiently during the on-state phases between off-state phases.

A SPICE model is built in section 4.4 to investigate the transient behavior and efficiency of the inverter. Additionally, scaling laws towards higher voltage and current levels were established. These were used to conduct a step-by-step design of a demonstrator with 10 V and 10 kA. The required superconducting tape length depends on:

Target Current Capacity

The analysis showed that the amount of parallel tapes needed for the required current capacity is the major contributor to the total length. More parallel tapes mean longer tape lengths exposed to a magnetic field, which are needed to build up a higher off-state resistance with each tape. Increasing the current density would reduce the number of required parallel tapes.

Target Efficiency

Also, the efficiency plays a significant role, as it directly impacts the maximum allowable leakage current. This determines the required off-state resistance based on the desired output parameters. For example, reducing the target efficiency from 99.9 % to 99 % results in a tenfold reduction in tape length.

Off-state Resistance

The number of parallel tapes and the required off-state resistance, needed to achieve the target efficiency, determine the necessary length of each superconducting tape in parallel. This length is governed by the resistance that each tape develops when exposed to a specific alternating magnetic field. A higher resistance per unit length means less superconducting material is needed overall. One possible method to increase this resistance is to apply a magnetic field with a higher amplitude or frequency.

Regarding the demonstrator presented in this work, this corresponds to a required tape length of 100.56 km to achieve an efficiency of 99.9 %. Reducing the target efficiency to 99.9 %, decreases the required length to 10.06 km. To further reduce the tape length, wider tapes were investigated and a four times wider superconductor is assumed. This would reduce the required length from 100.56 km to 7.35 km.

The implementation of HTS thin films on sapphire substrates was also examined more closely. These films have the potential to achieve higher total resistances by substituting the otherwise normal conducting substrate layer with an insulator. However, due to the lower critical current, more parallel conductors were required and led to an increased total amount of superconducting material.

Lastly, a method to generate long length magnetic fields based on the rectangular disk-up-down assembly was presented. An assembly with ten HTS tapes was manufactured and tested. The magnetization curve showed a maximum magnetic field of 109 mT at a current of 100 A. This method provides easy scalability to towards longer lengths while maintaining a small footprint.

5 Summary, Conclusion, and Outlook

In this work, a new potential application of high-temperature superconductors is investigated, and its feasibility is assessed, namely as a switch for power electronics. The main objective is to design and test a fully superconducting H-bridge inverter as a proof of concept. The switches in the circuit are based on the unique property of dynamic resistance that occurs when the superconductor is subjected to an alternating magnetic field. This effect is used to reach the off-state of the superconducting switch, while in on-state the switch is fully superconducting without any transport losses.

In chapter 2 an introduction to second generation high-temperature superconductors (HTS) used in this work is given. The limitation by temperature, magnetic field, and current under which the material is superconducting is discussed and additionally the behavior when the limits are exceeded. Moreover, the general layer structure and the function of each individual layer are described. The three possibilities of building up a resistance in the superconductor, which are already used for different applications, i.e. flux pump and fault current limiter, are presented. Furthermore, the method with the external alternating magnetic field in which dynamic resistance is generated and used in this work is examined in more detail. Analytical and numerical models are used to describe the dependency between the relevant variables, such as critical current I_c, magnetic field amplitude B_a and frequency f. These are compared with each other and the limits of validity of the individual methods are shown.

In order to fully understand the behavior of a superconducting switch and design an inverter based on these, preliminary experiments are required and given in chapter 3. The total and dynamic resistance are measured dependent on the external magnetic field amplitude and frequency for various combinations up to 277 mT and 1000 Hz. Values up to $120\,\mathrm{m\Omega\,m^{-1}}$ are achieved for the total resistance R_{tot}. These experimental results are compared with 2D FEM models. In the end, the measured resistances are higher than the analytical equations suggest. This is attributed to a thermal component. Both, the measurement results and the modeling, show that due to high ac losses, a temperature increase of up to 40 K is possible. This becomes even clearer with an additionally applied environmental protection by means of a polyimide laminate and results in higher resistances due to the poorer heat transfer. It is also shown that the total resistance of the switch can be further increased by removing the silver cap layer, whereby the conductor is still sufficiently stable and can exceed over 10 times the critical current criterion E_c of HTS materials. Different etching patterns are investigated.

To investigate the interaction between multiple superconducting switches, a test setup with two parallel connected switches is built and triggered in an alternating pattern. The current is completely distributed from one switch to the other, with this process occurring as an exponential increase or decrease. How quickly the current commutates depends on the dynamic resistance. The commutation times could be determined via an RL–circuit. A minimum time constant of 2.5 ms is achieved with excitation at 500 Hz and 337.68 mT.

With the information from previous chapters, a fully superconducting H–bridge inverter is designed and built in chapter 4. The inverter is based on four superconducting switches made from one slit HTS tape. This makes it possible to minimize the joint resistances between the switches. The switches are triggered in an alternating way, and it is shown that an dc input current of 50 A can be converted into an ac current in a rectangular profile which oscillates between −50 A and 50 A. This demonstrates the basic functionality of such a fully superconducting inverter. Again, the excitation parameters of the switches are varied to see the impact on the commutation times, and the methods of increasing the total resistance from chapter 3 are applied. Commutation times as low as 2.32 ms are achieved while suppressing the leakage current, which flows through the off-state switches, to 97.4 mA. To prove the thermal stability of the switches, a continuous switching test is conducted by varying the test period and the trigger pulse widths. The commutation times as an indicator for stability are analyzed over the whole test duration and shows consistent commutation times. Moreover, the variation of trigger pulse width from 1 s to 25 ms while keeping the same magnet excitation parameters leads to the same commutation times. Thus, pool boiling cooling is sufficient to cool the switches during the on-states.

A SPICE model is used to analyze the transient response and efficiency of the inverter, establishing scaling laws for higher voltages and currents. These scaling laws are applied to design a demonstrator with 10 V/10 kA. Such an inverter would require 100.56 km of HTS tape to achieve 99.9 % efficiency, with the number of parallel tapes as the main contributor to length. Reducing efficiency to 99 % or using four times wider tapes significantly decreases the required material down to 735 m. Alternative approaches, including HTS thin films on sapphire, are explored for higher off-state resistance but require more parallel conductors due to the constrained HTS layer thickness due to non-matching thermal expansion coefficients.

To magnetize long lengths of superconductors in a higher performance switch, a scalable method for generating long magnetic fields using, a rectangular disk-up-down assembly with HTS tapes is demonstrated, achieving a maximum magnetic field of 109 mT at 100 A with a compact setup.

In conclusion, this work provides the first steps towards demonstrating the feasibility of fully superconducting inverters as a new potential application. This involves a series of preliminary investigations that contribute to the understanding of such a circuit. An inverter is designed and its functionality demonstrated. Several approaches for scaling up to higher voltages and currents are also presented and investigated. While the fully superconducting inverter presented here is only a demonstrator as a proof of concept with a low-ohmic load, the results provide a foundation for future research and development towards more advanced systems. At present, the technology is at Technology Readiness Level (TRL) 3 to 4.

It is recommended that future work be focused on optimizing the switch design to further reduce losses, leakage currents, and to improve transient behavior by reaching higher off-state resistances and minimizing the inductance of the circuit. Some approaches have already been presented in sections 4.5.2 to 4.5.4. As an example, the rectangular disk-up-down assembly (rDUDA) stack for generating long length magnetic fields is investigated and tested. This method enables easy scaling to longer magnetic field lengths for higher off-state resistances. Further improvements to the manufacturing process are necessary. Improving the experimental testing to include calorimetric and electrical measurements will allow for a more accurate assessment of AC losses and overall efficiency, including switching losses during commutation and magnet losses. Additionally, to prove the scalability of the system, it requires simulation and building larger demonstrators, which could meet the standards of current semiconductor based inverters.

As demonstrated in section 4.5.2, wider HTS tapes are one possibility to reach higher currents while maintaining a reasonable amount of superconductors. Advances in HTS materials and manufacturing may lead to prototypes for common inverter applications in power grids or industrial power electronics.

Bibliography

[Ain18] Ainslie, Mark D; Bumby, Chris W; Jiang, Zhenan; Toyomoto, Ryuki and Amemiya, Naoyuki: "Numerical Modelling of Dynamic Resistance in High-Temperature Superconducting Coated-Conductor Wires". In: *Superconductor Science and Technology* 31.7 (July 1, 2018), p. 074003. doi: 10.1088/1361-6668/aac1d3.

[Arn21] Arndt, Tabea; Holzapfel, Bernhard; Noe, Mathias; Nast, Rainer; Hornung, Frank; Kläser, Marion and Kudymow, Andrej: "New Coil Configurations with 2G-HTS and Benefits for Applications". In: *Superconductor Science and Technology* 34.9 (Aug. 2021), p. 095006. doi: 10.1088/1361-6668/ac19f4.

[Bar57] Bardeen, J.; Cooper, L. N. and Schrieffer, J. R.: "Microscopic Theory of Superconductivity". In: *Physical Review* 106.1 (Apr. 1, 1957), pp. 162–164. doi: 10.1103/PhysRev.106.162.

[Bea64] Bean, Charles P.: "Magnetization of High-Field Superconductors". In: *Reviews of Modern Physics* 36.1 (Jan. 1, 1964), pp. 31–39. doi: 10.1103/RevModPhys.36.31.

[Bed86] Bednorz, J. G. and Müller, K. A.: "Possible highTc Superconductivity in the Ba–La–Cu–O System". In: *Zeitschrift für Physik B Condensed Matter* 64.2 (June 1, 1986), pp. 189–193. doi: 10.1007/BF01303701.

[Ber11] Berger, André: Entwicklung supraleitender, strombegrenzender Transformatoren. Karlsruher Schriftenreihe zur Supraleitung / Hrsg. Prof. Dr.-Ing. M. Noe, Prof. Dr. rer. nat. M. Siegel. Erscheinungsort nicht ermittelbar: KIT Scientific Publishing, 2011. 1 p.

[Bra93] Brandt, Ernst Helmut and Indenbom, Mikhail: "Type-II-superconductor Strip with Current in a Perpendicular Magnetic Field". In: *Physical Review B* 48.17 (Nov. 1, 1993), pp. 12893–12906. doi: 10.1103/PhysRevB.48.12893.

[Bro20] Brooks, J. M.; Ainslie, M. D.; Jiang, Zhenan; Pantoja, A. E.; Badcock, R. A. and Bumby, C. W.: "The Transient Voltage Response of ReBCO Coated Conductors Exhibiting Dynamic Resistance". In: *Superconductor Science and Technology* 33.3 (Jan. 1, 2020), p. 035007. doi: 10.1088/1361-6668/ab6bfe.

[Buc04] Buckel, Werner; Kleiner, Reinhold; Huebener, Rudolf and Buckel, Werner: Superconductivity: Fundamentals and Applications. 2., rev. and enl. ed. Physics Textbook. Weinheim: Wiley-VCH-Verl, 2004. 461 pp.

[Byk14] Bykovsky, N V; Fetisov, S S; Nosov, A A; Zubko, V V and Vysotsky, V S: "Analysis of Critical Current Reduction in Self-Field in Stacked Twisted 2G HTS Tapes". In: *Journal of Physics: Conference Series* 507.2 (May 2014), p. 022001. doi: 10.1088/1742-6596/507/2/022001.

[cera] ceraco ceramic coating GmbH: High Quality YBCO films and more - Home. High Quality YBCO films and more - Home. url: https://www.ceraco.de/ (visited on 04/21/2025).

[cerb] ceraco ceramic coating GmbH: High Quality YBCO films and more - Substrates. url: https://www.ceraco.de/ybco-films/substrates/ (visited on 04/21/2025).

[Cha11] Chang, K. S.; Jang, J. Y.; Lee, J. and Ko, T. K.: "An Experimental Analysis on the Design of HTS Current Lead with Respect to the Arrangement of Coated Conductor and the Shape of a Current Terminal". In: *Physica C: Superconductivity and its Applications*. The 23rd International Symposium on Superconductivity 471.21 (Nov. 1, 2011), pp. 1007–1011. doi: 10.1016/j.physc.2011.05.111.

[COM] COMSOL AB: Comsol Multiphysics®. Version 6.2. Stockholm, Sweden: COMSOL AB.

[Deu18] Deutscher, Guy: "High-Voltage Superconducting Fault Current Limiters Based on High-Diffusivity Dielectric Substrates". In: *Journal of Superconductivity and Novel Magnetism* 31.7 (July 1, 2018), pp. 1961–1963. doi: 10.1007/s10948-018-4633-8.

[dOli24] De Oliveira, Roberto A. H.; Taalibi, Othman and Arndt, Tabea: "Compact HTS Motor Designs Especially Useful to Be Cooled by Liquid Hydrogen". In: *IEEE Transactions on Applied Superconductivity* 34.6 (Sept. 2024), pp. 1–5. doi: 10.1109/TASC.2024.3392961.

[Dri12] Driessche, I. Van et al.: "Chemical Solution Deposition Using Ink-Jet Printing for YBCO Coated Conductors". In: *Superconductor Science and Technology* 25.6 (Apr. 2012), p. 065017. doi: 10.1088/0953-2048/25/6/065017.

[dSou21] De Sousa, Wescley Tiago Batista; Kottonau, Dustin; Karrari, Shahab; Geisbüsch, Jörn and Noe, Mathias: "Deployment of a Resistive Superconducting Fault Current Limiter for Improvement of Voltage Quality and Transient Recovery Voltage". In: *IEEE Transactions on Applied Superconductivity* 31.1 (Jan. 2021), pp. 1–9. doi: 10.1109/TASC.2020.3016460.

[dSou23] De Sousa, Wescley T. B.; Noe, Mathias; Huwer, Stefan and Reiser, Wolfgang: "Design of a 110-kV 2.0-kA SmartCoil Superconducting Fault Current Limiter". In: *IEEE Transactions on Applied Superconductivity* 33.4 (June 2023), pp. 1–9. doi: 10.1109/TASC.2023.3246818.

[Eki06] Ekin, Jack W.: Experimental Techniques for Low Temperature Measurements: Cryostat Design, Material Properties, and Superconductor Critical-Current Testing. Repr. Oxford: Oxford Univ. Press, 2006. 673 pp.

[Els12] Elschner, S. et al.: "ENSYSTROB – Design, Manufacturing and Test of a 3-Phase Resistive Fault Current Limiter Based on Coated Conductors for Medium Voltage Application". In: *Physica C: Superconductivity and its Applications.* 2011 Centennial Superconductivity Conference – EUCAS–ISEC–ICMC 482 (Nov. 20, 2012), pp. 98–104. doi: 10.1016/j.physc.2012.04.025.

[Gaw18] Gawith, J. D. D.; Geng, J.; Li, C.; Shen, B.; Zhang, X.; Ma, J. and Coombs, T. A.: "A Half-Bridge HTS Transformer–Rectifier Flux Pump with Two AC Field-Controlled Switches". In: *Superconductor Science and Technology* 31.8 (Jan. 1, 2018), p. 085002. doi: 10.1088/1361-6668/aac86d.

[Gaw19a] Gawith, J. D. D.; Ma, J.; Shen, B.; Li, C.; Yang, J.; Öztürk, Y. and Coombs, T. A.: "An HTS Power Switch Using YBCO Thin Film Controlled by AC Magnetic Field". In: *Superconductor Science and Technology* 32.9 (Jan. 1, 2019), p. 095007. doi: 10.1088/1361-6668/ab2d61.

[Gaw19b] Gawith, Jamie; Geng, Jianzhao; Ma, Jun; Shen, Boyang; Li, Chao and Coombs, Tim A.: "HTS Transformer–Rectifier Flux Pump Optimization". In: *IEEE Transactions on Applied Superconductivity* 29.5 (Jan. 1, 2019), pp. 1–5. doi: 10.1109/TASC.2019.2904444.

[Gen16] Geng, Jianzhao; Matsuda, K; Fu, Lin; Shen, Boyang; Zhang, Xiuchang and Coombs, T A: "Operational Research on a High- T_c Rectifier-Type Superconducting Flux Pump". In: *Superconductor Science and Technology* 29.3 (Mar. 1, 2016), p. 035015. doi: 10.1088/0953-2048/29/3/035015.

[Gen21] Geng, Jianzhao; Badcock, Rodney A. and Bumby, Chris W.: "A Wireless Rectifier for Inductively Energizing High Direct-Current High-Temperature Superconducting Magnets". In: *IEEE Transactions on Industrial Electronics* 68.4 (Jan. 1, 2021), pp. 3273–3281. doi: 10.1109/TIE.2020.2982095.

[Göm08] Gömöry, F; Šouc, J; Seiler, E; Vojenčiak, M and Granados, X: "Modification of Critical Current in HTSC Tape Conductors by a Ferromagnetic Layer". In: *Journal of Physics: Conference Series* 97 (Feb. 1, 2008), p. 012096. doi: 10.1088/1742-6596/97/1/012096.

[Goy04] Goyal, Amit; Paranthaman, M. Parans and Schoop, U.: "The RABiTS Approach: Using Rolling-Assisted Biaxially Textured Substrates for High-Performance YBCO Superconductors". In: *MRS Bulletin* 29.8 (Aug. 2004), pp. 552–561. doi: 10.1557/mrs2004.161.

[Gri04] Grilli, Francesco: Numerical Modelling of High Temperature Superconducting Tapes and Cables. 2909. Lausanne, EPFL, Jan. 1, 2004. doi: 10.5075/epfl-thesis-2909.

[Gri14a] Grilli, Francesco; Pardo, Enric; Stenvall, Antti; Nguyen, Doan N.; Weijia Yuan and Gomory, Fedor: "Computation of Losses in HTS Under the Action of Varying Magnetic Fields and Currents". In: *IEEE Transactions on Applied Superconductivity* 24.1 (Feb. 2014), pp. 78–110. doi: 10.1109/TASC.2013.2259827.

[Gri14b] Grilli, Francesco; Sirois, Frédéric; Zermeño, Victor M. R. and Vojenčiak, Michal: "Self-Consistent Modeling of the I_c of HTS Devices: How Accurate Do Models Really Need to Be?" In: *IEEE Transactions on Applied Superconductivity* 24.6 (Dec. 2014), pp. 1–8. doi: 10.1109/TASC.2014.2326925.

[Gro09] Grover, Frederick W.: Inductance Calculations: Working Formulas and Tables. Online-Ausg., unabridged republ. of the 2004 Dover ed. of the work 1. publ. in 1946, New York. Mineola, NY: Dover Publ, 2009. 1 p.

[Gro99] Gromoll, B. et al.: "Resistive Fault Current Limiters with YBCO Films 100 kVA Functional Model". In: *IEEE Transactions on Applied Superconductivity* 9.2 (June 1999), pp. 656–659. doi: 10.1109/77.783381.

[Guo13] Guo, L. S.; Chen, Y. Y.; Cheng, L.; Li, W.; Xiong, J.; Tao, B. W. and Yao, X.: "Liquid Phase Epitaxy of REBCO (RE=Y, Sm) Thick Films on YBCO Thin Film Deposited on LAO Substrate". In: *Journal of Crystal Growth* 366 (Mar. 1, 2013), pp. 47–50. doi: 10.1016/j.jcrysgro.2012.12.019.

[Har20] Harca, I M; Machaj, K; Tomkow, L and Glowacki, B A: "Magnetic Flux in Stacks of Superconducting Tapes of Different Architecture". In: *Superconductor Science and Technology* 33.11 (Sept. 2020), p. 115004. doi: 10.1088/1361-6668/abb11c.

[Hel14] Hellmann, Sebastian and Noe, Mathias: "Influence of Different Surface Treatments on the Heat Flux From Solids to Liquid Nitrogen". In: *IEEE Transactions on Applied Superconductivity* 24.3 (June 2014), pp. 1–5. doi: 10.1109/TASC.2013.2283772.

[Hel19] HELLMANN, Sebastian; ABPLANALP, Markus; ELSCHNER, Steffen; KUDYMOW, Andrej and NOE, Mathias: "Current Limitation Experiments on a 1 MVA-Class Superconducting Current Limiting Transformer". In: *IEEE Transactions on Applied Superconductivity* 29.5 (Aug. 2019), pp. 1–6. DOI: 10.1109/TASC.2019.2906804.

[Her16] HERZOG, F.; KUTZ,T. and STEMMLE, M.: Cooling Unit for the AmpaCity Project: Two Years Successful Operation. Kugel, T. International Institute of Refrigeration (IIR), 2016. DOI: 10.18462/IIR.ICCRT.2016.0029.

[Her19] HERZOG, F. and KUTZ, T.: Liquid Nitrogen Cooling System for Superconducting Electricity Cables – Experiences and Advantages. Version 1. International Institute of Refrigeration (IIR), 2019. DOI: 10.18462/IIR.CRYO.2019.0002.

[Hol25] HOLZAPFEL, Bernhard and BAGRETS, Nadezda: "Progress on ReBCO Tape Research at KIT". HFM Annual Meeting 2025 (Geneva). Feb. 12, 2025.

[IEC06a] IEC: Superconductivity - Part 3: Critical Current Measurement - DC Critical Current of Ag- and/or Ag Alloy-Sheathed Bi-2212 and Bi-2223 Oxide Superconductors. IEC 61788-3:2006. Version 2.0. 2006. Valid.

[IEC06b] IEC: Superconductivity – Part 10: Critical Temperature Measurement – Critical Temperature of Composite Superconductors by a Resistance Method. IEC 61788-10:2006. Version 2.0. 2006. Valid.

[IEC20] IEC: Superconductivity - Part 26: Critical Current Measurement - DC Critical Current of RE-Ba-Cu-O Composite Superconductors. IEC 61788-26:2020. 2020. valid.

[Iij00] IIJIMA, Yasuhiro and MATSUMOTO, Kaname: "High-Temperature-Superconductor Coated Conductors: Technical Progress in Japan". In: *Superconductor Science and Technology* 13.1 (Jan. 2000), p. 68. DOI: 10.1088/0953-2048/13/1/310.

[Jia17a] JIANG, Zhenan; TOYOMOTO, Ryuki; AMEMIYA, Naoyuki; BUMBY, Chris W.; BADCOCK, Rodney A. and LONG, Nicholas J.: "Dynamic Resistance Measurements in a GdBCO-Coated Conductor". In: *IEEE Transactions on Applied Superconductivity* 27.4 (Jan. 1, 2017), pp. 1–5. DOI: 10.1109/TASC.2016.2644107.

[Jia17b] JIANG, Zhenan; TOYOMOTO, Ryuki; AMEMIYA, Naoyuki; ZHANG, Xingyou and BUMBY, Chris W: "Dynamic Resistance of a High-Tc Coated Conductor Wire in a Perpendicular Magnetic Field at 77 K". In: *Superconductor Science and Technology* 30.3 (Jan. 2017), 03LT01. DOI: 10.1088/1361-6668/aa54e5.

[Jia18] JIANG, Z.; ZHOU, W.; BUMBY, C. W.; STAINES, M.; LI, Q.; BADCOCK, R. A.; LONG, N. J. and FANG, J.: "Dynamic Resistance Measurement of a Four-Tape YBCO Stack in a Perpendicular Magnetic Field". In: *IEEE Transactions on Applied Superconductivity* 28.4 (Jan. 1, 2018), pp. 1–5. DOI: 10.1109/TASC.2017.2787178.

[Kha24] Khallouq, Keltoum: Exploring High-Temperature Superconductivity in the YBCO System: From Theory to Experiments. SpringerBriefs in Materials. Cham: Springer Nature Switzerland, 2024. doi: 10.1007/978-3-031-66238-6.

[Kim13] Kim, Youngjae; Bascuñán, Juan; Lecrevisse, Thibault; Hahn, Seungyong; Voccio, John; Park, Dong Keun and Iwasa, Yukikazu: "YBCO and Bi2223 Coils for High Field LTS/HTS NMR Magnets: HTS-HTS Joint Resistivity". In: *IEEE Transactions on Applied Superconductivity* 23.3 (June 2013), pp. 6800704–6800704. doi: 10.1109/TASC.2013.2243195.

[Kim15] Kim, S. B.; Takahashi, M.; Saito, R.; Park, Y. J.; Lee, M. W.; Oh, Y. K. and Ann, H. S.: "Current Bypassing Properties by Thermal Switch for PCS Application on NMR/MRI HTS Magnets". In: *Physics Procedia.* Proceedings of the 27th International Symposium on Superconductivity (ISS 2014) November 25-27, 2014, Tokyo, Japan 65 (Jan. 1, 2015), pp. 149–152. doi: 10.1016/j.phpro.2015.05.088.

[Kom95] Komarek, Peter: Hochstromanwendung der Supraleitung. Teubner Studienbücher Angewandte Physik. Stuttgart: Teubner, 1995.

[Kra03] Kraemer, H.-P.; Schmidt, W.; Utz, B. and Neumueller, H.-W.: "Switching Behavior of YBCO Thin Film Conductors in Resistive Fault Current Limiters". In: *IEEE Transactions on Applied Superconductivity* 13 (2 II 2003), pp. 2044–2047. doi: 10.1109/TASC.2003.812980.

[Kud07] Kudymow, A.; Noe, M.; Schacherer, C.; Kinder, H. and Prusseit, W.: "Investigation of YBCO Coated Conductor for Application in Resistive Superconducting Fault Current Limiters". In: *IEEE Transactions on Applied Superconductivity* 17.2 (June 2007), pp. 3499–3502. doi: 10.1109/TASC.2007.899578.

[Léc15] Lécrevisse, Thibault; Bascuñán, Juan; Hahn, Seungyong; Kim, Youngjae; Song, Jungbin and Iwasa, Yukikazu: "Tape-to-Tape Joint Resistances of a Magnet Assembled From (RE)BCO Double-Pancake Coils". In: *IEEE Transactions on Applied Superconductivity* 25.3 (June 2015), pp. 1–5. doi: 10.1109/TASC.2014.2372891.

[Lee23] Lee, Hunju: "REBCO Tape at SuNAM: State of the Art and Prospects". 1st High Temperature Superconductors for Accelerator Technology (HiTAT) Workshop (CERN). Mar. 9, 2023.

[LEM25] LEM: Hall Effect Sensor | Hall Effect Current and Voltage Sensors. 2025. url: https://www.lem.com/en/hall-effect-current-sensors (visited on 02/17/2025).

[Leu22] LEUW, Bradley; GENG, Jianzhao; RICE, James H P; MOSELEY, Dominic A and BADCOCK, Rodney A: "A Half-Wave Superconducting Transformer-Rectifier Flux Pump Using J $_c$ (B) Switches". In: *Superconductor Science and Technology* 35.3 (Mar. 1, 2022), p. 035009. DOI: 10.1088/1361-6668/ac4f3d.

[Li02] LI, M.; MA, B.; KORITALA, R. E.; FISHER, B. L.; VENKATARAMAN, K. and BALACHANDRAN, U.: "Pulsed Laser Deposition of YBCO Thin Films on IBAD–YSZ Substrates". In: *Superconductor Science and Technology* 16.1 (Dec. 2002), p. 105. DOI: 10.1088/0953-2048/16/1/319.

[Li18] LI, Chao; GENG, Jianzhao; GAWITH, Jamie; SHEN, Boyang; ZHANG, Xiuchang; ZHANG, Heng; MA, Jun and COOMBS, Tim A.: "Design for a Persistent Current Switch Controlled by Alternating Current Magnetic Field". In: *IEEE Transactions on Applied Superconductivity* 28.4 (Jan. 1, 2018), pp. 1–5. DOI: 10.1109/TASC.2018.2809545.

[Li19] LI, Chao; GENG, Jianzhao; SHEN, Boyang; LI, Xinru; GAWITH, Jamie; MA, Jun; YANG, Jiabin and COOMBS, T. A.: "Persistent Current Switch for HTS Superconducting Magnets: Design, Control Strategy, and Test Results". In: *IEEE Transactions on Applied Superconductivity* 29.2 (Jan. 1, 2019), pp. 1–4. DOI: 10.1109/TASC.2018.2889463.

[Li24] LI, Weiyong; ZHOU, Yong; WANG, Yongmao; LENG, Feng; ZHAO, Yutian; HE, Zehao; YU, Shuai; ZHENG, Jun and NIU, Xiaojun: "Performance of a Persistent Current Switch for Large-Scale HTS Magnet". In: *IEEE Transactions on Applied Superconductivity* 34.8 (Nov. 2024), pp. 1–4. DOI: 10.1109/TASC.2024.3420319.

[Lia17] LIANG, Fei; VENUTURUMILLI, Sriharsha; ZHANG, Huiming; ZHANG, Min; KVITKOVIC, Jozef; PAMIDI, Sastry; WANG, Yawei and YUAN, Weijia: "A Finite Element Model for Simulating Second Generation High Temperature Superconducting Coils/Stacks with Large Number of Turns". In: *Journal of Applied Physics* 122.4 (July 27, 2017), p. 043903. DOI: 10.1063/1.4995802.

[Liu22] LIU, Wei; WANG, Yinshun; WANG, Jian; ZHANG, Guangyi; LIU, Yating; CHAI, Hua and WANG, Jiawen: "Intrinsic Influence of Dynamic Resistance on Current Distribution of Two HTS Parallel Branches Without Terminal Contact Resistance". In: *IEEE Transactions on Applied Superconductivity* 32.1 (Jan. 2022), pp. 1–5. DOI: 10.1109/TASC.2021.3126326.

[Lu08] LU, J.; CHOI, E. S. and ZHOU, H. D.: "Physical Properties of Hastelloy ® C-276™ at Cryogenic Temperatures". In: *Journal of Applied Physics* 103.6 (Jan. 1, 2008), p. 064908. DOI: 10.1063/1.2899058.

[Maj15] Majkic, Goran; Galstyan, Eduard and Selvamanickam, Venkat: "High Performance 2G-HTS Wire Using a Novel MOCVD System". In: *IEEE Transactions on Applied Superconductivity* 25.3 (June 2015), pp. 1–4. doi: 10.1109/TASC.2014.2372902.

[Mat79] Matula, R. A.: "Electrical Resistivity of Copper, Gold, Palladium, and Silver". In: *Journal of Physical and Chemical Reference Data* 8.4 (Jan. 1, 1979), pp. 1147–1298. doi: 10.1063/1.555614.

[Mel24] Mellerud, Runar; Hartmann, Christian; Klop, Casper Leonard and Nøland, Jonas Kristiansen: "Influence of Eddy Current Losses in Nonsuperconducting Layers of HTS in Superconducting Electrical Machines". In: *IEEE Transactions on Applied Superconductivity* 34.7 (Oct. 2024), pp. 1–7. doi: 10.1109/TASC.2024.3431232.

[Mer64] Merte Jr., H. and Clark, J. A.: "Boiling Heat Transfer With Cryogenic Fluids at Standard, Fractional, and Near-Zero Gravity". In: *Journal of Heat Transfer* 86.3 (Aug. 1, 1964), pp. 351–358. doi: 10.1115/1.3688689.

[Mic17] Michael, Philip C.; Qu, Timing; Voccio, John; Bascuñán, Juan; Hahn, Seungyong and Iwasa, Yukikazu: "A REBCO Persistent-Current Switch (PCS): Test Results and Switch Heater Performance". In: *IEEE Transactions on Applied Superconductivity* 27.4 (June 2017), pp. 1–5. doi: 10.1109/TASC.2017.2652303.

[Mik01] Mikitik, Grigorii P. and Brandt, Ernst Helmut: "Generation of a Dc Voltage by an Ac Magnetic Field in Type-II Superconductors". In: *Physical Review B* 64.9 (Aug. 1, 2001), p. 092502. doi: 10.1103/PhysRevB.64.092502.

[Mol25] Molodyk, Alexander: "Advances and Perspectives at Faraday Factory Japan". Coated Conductor for Applications (CCA) Workshop 2025 (Geneva). Mar. 11, 2025.

[Mos17] Mosser, Vincent; Matringe, Nicolas and Haddab, Youcef: "A Spinning Current Circuit for Hall Measurements Down to the Nanotesla Range". In: *IEEE Transactions on Instrumentation and Measurement* 66.4 (Apr. 2017), pp. 637–650. doi: 10.1109/TIM.2017.2649858.

[Mun90] Munter, P. J. A.: "A Low-Offset Spinning-Current Hall Plate". In: *Sensors and Actuators A: Physical* 22.1 (June 1, 1990), pp. 743–746. doi: 10.1016/0924-4247(89)80069-X.

[NI] NI: Instrument Bus Performance – Making Sense of Competing Bus Technologies for Instrument Control. url: https://www.ni.com/en/support/downloads/instrument-drivers/instrument-bus-performance---making-sense-of-competing-bus-techn.html (visited on 04/07/2025).

[NI25] NI: Ni/Nidaqmx-Python. Version 0.8.0. NI, Jan. 31, 2025.

[Nib01] NIBBIO, N.; STAVREV, S. and DUTOIT, B.: "Finite Element Method Simulation of AC Loss in HTS Tapes with B-dependent E-J Power Law". In: *IEEE Transactions on Appiled Superconductivity* 11.1 (Mar. 2001), pp. 2631–2634. DOI: 10.1109/77.920408.

[Noe07] NOE, Mathias and STEURER, Michael: "High-Temperature Superconductor Fault Current Limiters: Concepts, Applications, and Development Status". In: *Superconductor Science and Technology* 20.3 (Jan. 2007), R15. DOI: 10.1088/0953-2048/20/3/R01.

[Noe23] NOE, Mathias et al.: 380 kV Superconducting Fault Current Limiter Feasibility Study. KIT Scientific Publishing, 2023. DOI: 10.5445/KSP/1000161057.

[Onn91] ONNES, H. Kamerlingh: "Further Experiments with Liquid Helium. G. On the Electrical Resistance of Pure Metals, Etc. VI. On the Sudden Change in the Rate at Which the Resistance of Mercury Disappears." In: *Through Measurement to Knowledge: The Selected Papers of Heike Kamerlingh Onnes 1853–1926*. Ed. by ONNES, Heike Kamerlingh; GAVROGLU, Kostas and GOUDAROULIS, Yorgos. Dordrecht: Springer Netherlands, 1991, pp. 267–272. DOI: 10.1007/978-94-009-2079-8_17.

[Oom00] OOMEN, M.P: AC Loss in Superconducting Tapes and Cables. 1e uitg. Enschede: Vakgroep Lage Temperaturen, Jan. 1, 2000. 176 pp.

[Oom05] OOMEN, M.P.; LEGHISSA, M.; RIES, G.; PROELSS, N.; NEUMUELLER, H.-W.; STEINMEYER, F.; VESTER, M. and DAVIES, F.: "HTS Flux Pump for Cryogen-Free HTS Magnets". In: *IEEE Transactions on Applied Superconductivity* 15 (2 PART II 2005), pp. 1465–1468. DOI: 10.1109/TASC.2005.849129.

[Oom99] OOMEN, M. P.; RIEGER, J.; LEGHISSA, M.; HAKEN, B. and KATE, H. H. J.: "Dynamic Resistance in a Slab-like Superconductor with J c (B) Dependence". In: *Superconductor Science and Technology* 12.6 (Jan. 1, 1999), pp. 382–387. DOI: 10.1088/0953-2048/12/6/309.

[Ott16] OTTEN, Simon; KARIO, Anna; KLING, Andrea and GOLDACKER, Wilfried: "Bending Properties of Different REBCO Coated Conductor Tapes and Roebel Cables at T = 77 K". In: *Superconductor Science and Technology* 29.12 (Dec. 1, 2016), p. 125003. DOI: 10.1088/0953-2048/29/12/125003.

[Pha22] PHAM, Quoc Hung and NOE, Mathias: "Experimental Investigation of the Switching Behavior of High-Temperature Superconductors with an Alternating Magnetic Field". In: Applied Superconductivity Conference (ASC) 2022. Honolulu, HI, USA, 2022. DOI: 10.5445/IR/1000177221.

[Pha25a] Pham, Quoc Hung; Nast, Rainer and Noe, Mathias: "Resistance of High-Temperature Superconducting Tapes Triggered by Alternating Magnetic Field". In: *IEEE Transactions on Applied Superconductivity* 35.3 (May 2025), pp. 1–8. doi: 10.1109/TASC.2025.3550837.

[Pha25b] Pham, Quoc Hung; Noe, Mathias and Kudymow, Andrej: "Magnetic Field Triggered Switching of High-Temperature Superconductors—Basic Experiments". In: *IEEE Transactions on Applied Superconductivity* 35.2 (Mar. 2025), pp. 1–7. doi: 10.1109/TASC.2024.3514180.

[Phi96] Phillips, Julia M.: "Substrate Selection for High-temperature Superconducting Thin Films". In: *Journal of Applied Physics* 79.4 (Feb. 15, 1996), pp. 1829–1848. doi: 10.1063/1.362675.

[Pią12] Piątek, Z.; Baron, B.; Szczegielniak, T.; Kusiak, D. and Pasierbek, A.: "Self Inductance of Long Conductor of Rectangular Cross Section". In: *PRZEGLĄD ELEKTROTECHNICZNY* 88.8 (2012), pp. 323–326.

[Pop22] Pop, C et al.: "High Critical Current Solution Derived YBa2Cu3O7 Films Grown on Sapphire". In: *Superconductor Science and Technology* 35.5 (Apr. 2022), p. 054007. doi: 10.1088/1361-6668/ac5be9.

[Pyt] Python Software Foundation: What's New In Python 3.11. Python documentation. url: https://docs.python.org/3/whatsnew/3.11.html (visited on 04/07/2025).

[Roh] Rohde & Schwarz GmbH & Co. KG: Connectivity of Rohde&Schwarz Signal Generators. url: https://www.rohde-schwarz.com/cac/applications/connectivity-of-rohde-schwarz-signal-generators-application-note_56280-15626.html (visited on 04/07/2025).

[Ros08] Rosa, E. B.: "The Self and Mutual-Inductances of Linear Conductors". In: *Bulletin of the Bureau of Standards* 4.2 (Jan. 1, 1908), p. 301. doi: 10.6028/bulletin.088.

[San10] Sanchez, Alvaro; Del-Valle, Nuria; Navau, Carles and Chen, Du-Xing: "Influence of Magnetic Substrate in the Transport Critical Current of Superconducting Tapes". In: *Applied Physics Letters* 97.7 (Aug. 17, 2010), p. 072504. doi: 10.1063/1.3481082.

[San23] Sandra, Jithin Sai; Paidpilli, Mahesh; Sun, Sicong; Suarez-Villagran, Martha; Yerraguravagari, Vamsi; Wosik, Jarek and Selvamanickam, Venkat: "Applications-Oriented Development of Buffer Architecture for REBCO Films on Nonmetallic Substrates". In: *IEEE Transactions on Applied Superconductivity* 33.7 (Oct. 2023), pp. 1–6. doi: 10.1109/TASC.2023.3289651.

[Sch07] Schott, Christian; Racz, Robert; Manco, Angelo and Simonne, Nicolas: "CMOS Single-Chip Electronic Compass With Microcontroller". In: *IEEE Journal of Solid-State Circuits* 42.12 (Dec. 2007), pp. 2923–2933. doi: 10.1109/JSSC.2007.908694.

[Sch09] Schacherer, Christian: Theoretische Und Experimentelle Untersuchungen Zur Entwicklung Supraleitender Resistiver Strombegrenzer. KIT Scientific Publishing, Sept. 25, 2009. doi: 10.5445/KSP/1000012282.

[Sch17] Schacherer, Christian; Bauer, Anne; Elschner, Steffen; Goldacker, Wilfried; Kraemer, Hans-Peter; Kudymow, Andrej; Naeckel, Oliver; Strauss, Severin and Zermeno, Victor M. R.: "SmartCoil - Concept of a Full-Scale Demonstrator of a Shielded Core Type Superconducting Fault Current Limiter". In: *IEEE Transactions on Applied Superconductivity* 27.4 (June 2017), pp. 1–5. doi: 10.1109/TASC.2016.2642139.

[Sei15] Seidel, Paul, ed.: Applied Superconductivity: Handbook on Devices and Applications. 1st ed. Wiley, Jan. 7, 2015. doi: 10.1002/9783527670635.

[Sel09] Selvamanickam, Venkat et al.: "High Performance 2G Wires: From R&D to Pilot-Scale Manufacturing". In: *IEEE Transactions on Applied Superconductivity* 19.3 (June 2009), pp. 3225–3230. doi: 10.1109/TASC.2009.2018792.

[Sem02] Semerad, R; Knauf, J; Irgmaier, K and Prusseit, W: "RE-123 Thin Films for Microwave Applications". In: *Physica C: Superconductivity* 378–381 (Oct. 1, 2002), pp. 1414–1418. doi: 10.1016/S0921-4534(02)01735-5.

[Sha21] Sharma, R.G.: Superconductivity: Basics and Applications to Magnets. Vol. 214. Springer Series in Materials Science. Cham: Springer International Publishing, 2021. doi: 10.1007/978-3-030-75672-7.

[She20] Shen, Boyang; Grilli, Francesco and Coombs, Tim: "Overview of H-Formulation: A Versatile Tool for Modeling Electromagnetics in High-Temperature Superconductor Applications". In: *IEEE Access* 8 (2020), pp. 100403–100414. doi: 10.1109/ACCESS.2020.2996177.

[She93] Shevchenko, O.A.; ten Kate, H.H.J.; Krooshoop, H.J.G.; Markovsky, N.V. and Mulder, G.B.J.: "Development of a 50-60 Hz Thermally Switched Superconducting Rectifier". In: *IEEE Transactions on Applied Superconductivity* 3.1 (Mar. 1993), pp. 590–593. doi: 10.1109/77.233774.

[Shi22] Shiva Kumar, S. and Lim, Hankwon: "An Overview of Water Electrolysis Technologies for Green Hydrogen Production". In: *Energy Reports* 8 (Nov. 1, 2022), pp. 13793–13813. doi: 10.1016/j.egyr.2022.10.127.

[Smi94] Smith, J. L. et al.: "Low-Temperature Critical Field of YBCO". In: *Journal of Superconductivity* 7.2 (Apr. 1, 1994), pp. 269–270. doi: 10.1007/BF00724550.

[Son25] Song, Bai: "Conductor Manufacturing: Advances and Perspectives @SST". Coated Conductor for Applications (CCA) Workshop 2025 (Geneva). Mar. 11, 2025.

[Ste16] Stemmle, M.; Merschel, F. and Noe, M.: "Ampacity Project - World's First Superconducting Cable and Fault Current Limiter Installation in a German City Center". In: *Research, Fabrication and Applications of Bi-2223 HTS Wires.* 2016, pp. 263–278. doi: 10.1142/9789814749268_0019.

[Str19] Strickland, Nicholas M.; Wimbush, Stuart C.; Rupich, Martin W. and Long, Nicholas J.: "Asymmetries in the Field and Angle Dependences of the Critical Current in HTS Tapes". In: *IEEE Transactions on Applied Superconductivity* 29.5 (Aug. 2019), pp. 1–4. doi: 10.1109/TASC.2019.2894278.

[Tab] Tabor Electronics Ltd.: Datasheet Tabor WW5064.

[Tex] Texas Instruments: DRV411 Data Sheet, Product Information and Support | TI.Com. url: https://www.ti.com/product/DRV411#pps (visited on 02/17/2025).

[The25] Thevalappilly Paulose, Manoj; Sandra, Jithin Sai; Sayeed, Md Abu and Selvamanickam, Venkat: "Development of REBCO Thin Films Using MOCVD on Non-Standard Buffers and Substrates". In: *IEEE Transactions on Applied Superconductivity* 35.5 (Aug. 2025), pp. 1–5. doi: 10.1109/TASC.2024.3513942.

[Tix23] Tixador, Pascal: "Fault Current Limiter Based on High Temperature Superconductors". In: *Physica C: Superconductivity and its Applications* 615 (Dec. 15, 2023), p. 1354398. doi: 10.1016/j.physc.2023.1354398.

[tKat81] Ten Kate, H. H. J.; Bunk, P. B.; Britton, R. B. and van de Klundert, L. J. M.: "High Current and High Power Superconducting Rectifiers". In: *Cryogenics* 21.5 (May 1, 1981), pp. 291–296. doi: 10.1016/0011-2275(81)90006-0.

[Tos04] Tosaka, T.; Kuriyama, T.; Yamaji, M.; Kuwano, K.; Igarashi, M. and Terai, M.: "Development of a Persistent Current Switch for HTS Magnets". In: *IEEE Transactions on Applied Superconductivity* 14.2 (Jan. 1, 2004), pp. 1218–1221. doi: 10.1109/TASC.2004.830534.

[Uso17] Usoskin, A; Rao, V V; Dietrich, R; Rutt, A and Schlenga, K: "Fault Current Limiters and Fault Current Switches Based on Wide HTS Tapes: Low Cryo-Consumption, New Applications". In: *IOP Conference Series: Materials Science and Engineering* 171 (Feb. 2017), p. 012119. doi: 10.1088/1757-899X/171/1/012119.

[VAC25] VACUUMSCHMELZE GmbH & Co. KG: Closed Loop Current Sensors | VAC. 2025. URL: https://www.vacuumschmelze.com/products/current-sensors/closed-loop-current-sensors (visited on 02/17/2025).

[vdKlu81a] Van de Klundert, L. J. M. and ten Kate, H. H. J.: "On Fully Superconducting Rectifiers and Fluxpumps. A Review. Part 2: Commutation Modes, Characteristics and Switches". In: *Cryogenics* 21.5 (May 1, 1981), pp. 267–277. doi: 10.1016/0011-2275(81)90002-3.

[vdKlu81b] Van de Klundert, L.J.M. and Ten Kate, H.H.J.: "Fully Superconducting Rectifiers and Fluxpumps Part 1: Realized Methods for Pumping Flux". In: *Cryogenics* 21.4 (Apr. 1981), pp. 195–206. doi: 10.1016/0011-2275(81)90195-8.

[Wan22] Wang, Wei; Wei, Jiafu; Yang, Chao; Wu, Chenghuai and Li, Hong: "Review of High Temperature Superconducting Flux Pumps". In: *Superconductivity* 3 (Sept. 1, 2022), p. 100022. doi: 10.1016/j.supcon.2022.100022.

[Wan23] Wang, Rui; Liu, Yingzhen; Cao, Jiwei; Li, Liyi; Liu, Xiaokun; Xue, Haida and Arndt, Tabea: "Preliminary Design Optimization of a Fully Superconducting Motor Based on Disk-up-down-Assembly Magnets". In: *Superconductor Science and Technology* 36.5 (Apr. 2023), p. 054003. doi: 10.1088/1361-6668/acc822.

[Wil15] Willering, G P; Van Der Laan, D C; Weijers, H W; Noyes, P D; Miller, G E and Viouchkov, Y: "Effect of Variations in Terminal Contact Resistances on the Current Distribution in High-Temperature Superconducting Cables". In: *Superconductor Science and Technology* 28.3 (Mar. 1, 2015), p. 035001. doi: 10.1088/0953-2048/28/3/035001.

[Wil86] Wilson, Martin N.: Superconducting Magnets. Reprinted. Monographs on Cryogenics 2. Oxford: Clarendon Press, 1986. 335 pp.

[Wim17a] Wimbush, Stuart; Strickland, Nick and Pantoja, Andres E.: Critical Current Characterisation of SuperPower Advanced Pinning 2G HTS Superconducting Wire. figshare, Jan. 11, 2017. doi: 10.6084/m9.figshare.4256624.v3.

[Wim17b] Wimbush, Stuart C. and Strickland, Nicholas M.: "A Public Database of High-Temperature Superconductor Critical Current Data". In: *IEEE Transactions on Applied Superconductivity* 27.4 (June 2017), pp. 1–5. doi: 10.1109/TASC.2016.2628700.

[Wim19] Wimbush, Stuart and Strickland, Nick: Critical Current Characterisation of THEVA Pro-Line 2G HTS Superconducting Wire. figshare, 2019. doi: 10.6084/M9.FIGSHARE.3759327.V2.

[Wim21] Wimbush, Stuart; Strickland, Nick and Pantoja, Andres: Critical Current Characterisation of Fujikura FESC 2G HTS Superconducting Wire. figshare, 2021. doi: 10.6084/M9.FIGSHARE.15095703.V1.

[Wim22] Wimbush, Stuart; Strickland, Nick and Pantoja, Andres: Critical Current Characterisation of Shanghai Superconductor Low Field High Temperature 2G HTS Superconducting Wire. figshare, 2022. doi: 10.6084/M9.FIGSHARE.19185092.V1.

[Zha19a] Zhang, Hongye; Yao, Min; Jiang, Zhenan; Xin, Ying and Li, Quan: "Dependence of Dynamic Loss on Critical Current and n -Value of HTS Coated Conductors". In: *IEEE Transactions on Applied Superconductivity* 29.8 (Dec. 2019), pp. 1–7. doi: 10.1109/TASC.2019.2948993.

[Zha19b] Zhang, Yuan; Sun, Sicong; Pratap, Rudra; Galstyan, Eduard; Wosik, Jarek and Selvamanickam, Venkat: "Development of REBCO Tapes on Nonmetallic Flexible Substrates for RF Applications". In: *IEEE Transactions on Applied Superconductivity* 29.5 (Aug. 2019), pp. 1–5. doi: 10.1109/TASC.2019.2896545.

[Zha20a] Zhang, Hongye; Chen, Hongyi; Jiang, Zhenan; Yang, Tianhui; Xin, Ying; Mueller, Markus and Li, Quan: "A Full-Range Formulation for Dynamic Loss of High-Temperature Superconductor Coated Conductors". In: *Superconductor Science and Technology* 33.5 (May 1, 2020), 05LT01. doi: 10.1088/1361-6668/ab7b0d.

[Zha20b] Zhang, Hongye; Yao, Min; Kails, Kevin; Machura, Philip; Mueller, Markus; Jiang, Zhenan; Xin, Ying and Li, Quan: "Modelling of Electromagnetic Loss in HTS Coated Conductors over a Wide Frequency Band". In: *Superconductor Science and Technology* 33.2 (Jan. 1, 2020), p. 025004. doi: 10.1088/1361-6668/ab6022.

[Zha22] Zhang, Hongye; Shen, Boyang; Chen, Xiaoyuan and Jiang, Zhenan: "Dynamic Resistance and Dynamic Loss in a ReBCO Superconductor". In: *Superconductor Science and Technology* 35.11 (Nov. 1, 2022), p. 113001. doi: 10.1088/1361-6668/ac95d5.

[Zhu21] Zhu, Lingfeng; Wang, Yinshun; Liu, Wei; Liu, Yating; Nie, Yang and Pi, Wei: "Conceptual Design of HTS Bitter Magnet Above 25T Using a Fast Magnetic Field Computational Method". In: *IEEE Transactions on Applied Superconductivity* 31.5 (Aug. 2021), pp. 1–6. doi: 10.1109/TASC.2021.3068370.

Own Publications

[1] PHAM, Quoc Hung; NAST, Rainer and NOE, Mathias: "Resistance of High-Temperature Superconducting Tapes Triggered by Alternating Magnetic Field". In: *IEEE Transactions on Applied Superconductivity* 35.3 (May 2025), pp. 1–8. DOI: 10.1109/TASC.2025.3550837.

[2] PHAM, Quoc Hung and NOE, Mathias: "Basic Experiments With a Superconducting Full-Bridge Current-Driven Inverter". In: *IEEE Transactions on Applied Superconductivity* 35.4 (June 2025), pp. 1–6. DOI: 10.1109/TASC.2025.3556394.

[3] PHAM, Quoc Hung; NOE, Mathias and KUDYMOW, Andrej: "Magnetic Field Triggered Switching of High-Temperature Superconductors—Basic Experiments". In: *IEEE Transactions on Applied Superconductivity* 35.2 (Mar. 2025), pp. 1–7. DOI: 10.1109/TASC.2024.3514180.

Patents

[1] HUWER, Stefan; PHAM, Quoc Hung; REISER, Dr Wolfgang; MARTZ, Simon and NOE, Dr Mathias: "Hochtemperatur supraleitende Schaltvorrichtung". Pat. LU502329B1 (LU). VISION ELECTRIC SUPER CONDUCTORS GMBH, KARLSRUHER INSTITUT FUER TECHNOLOGIE (KIT). Dec. 22, 2023.

[2] NOE, Dr Mathias; REISER, Dr Wolfgang; PHAM, Quoc Hung; MARTZ, Simon and HUWER, Stefan: "Hochtemperatur supraleitende Schaltvorrichtung". European pat. 4297278A1. VISION ELECTRIC SUPER CONDUCTORS GMBH, KARLSRUHER INSTITUT FUER TECHNOLOGIE (KIT). Dec. 27, 2023.

Conference Contributions

[1] PHAM, Quoc Hung and NOE, Mathias: "Experimental Investigation of the Switching Behavior of High-Temperature Superconductors with an Alternating Magnetic Field". In: Applied Superconductivity Conference (ASC) 2022. Honolulu, HI, USA, 2022. DOI: 10.5445/IR/1000177221.

[2] PHAM, Quoc Hung and NOE, Mathias: "Switching Device with Magnet Field-Triggered High-Temperature Superconducting Switching Units". In: European Conference on

Applied Superconductivity (EUCAS) 2023. Bologna, Italy, 2023. DOI: 10 . 5445 / IR / 1000177219.

[3] Noe, Mathias and Pham, Quoc Hung: “Experiments with a Superconducting Full Bridge Inverter”. In: Applied Superconductivity Conference (ASC) 2024. Salt Lake City, UT, USA, 2024. DOI: 10.5445/IR/1000175489.

List of Figures

List of Tables

Acronyms

2D	two-dimensional
ac	alternate current
Ag	silver
Au	gold
BCS	Bardeen–Cooper–Schrieffer
CSD	chemical solution deposition
dc	direct current
FEM	finite element method
FFT	fast Fourier transform
GPIB	General Purpose Interface Bus
HTS	high-temperature superconductor
IC	integrated circuit
LHe	liquid helium
LN_2	liquid nitrogen
LTS	low-temperature superconductor
MOCVD	metal–oxide chemical vapor deposition

MRI magnetic resonance imaging

NMR nuclear magnetic resonance

PCB printed circuit board

PVD pulsed vapor deposition

rDUDA rectangular disk-up-down assembly

ReBCO rare-earth barium copper oxide

RFCL resistive fault current limiter

RT room temperature

SEM scanning electron microscope

SMES superconducting magnetic energy storage

SPICE Simulation Program with Integrated Circuit Emphasis

TRL Technology Readiness Level

VISA virtual instrument software architecture

YBCO yttrium barium copper oxide

Appendix

A Current Transducer Electronics

The printed circuit board (PCB) in fig. A.1 is used to drive the closed-loop hall effect current transducers and converts the primary current to a measurable voltage using the IC DRV411 by Texas Instruments. Figure A.2 shows the schematic and table A.1 the bill of materials of the PCB.

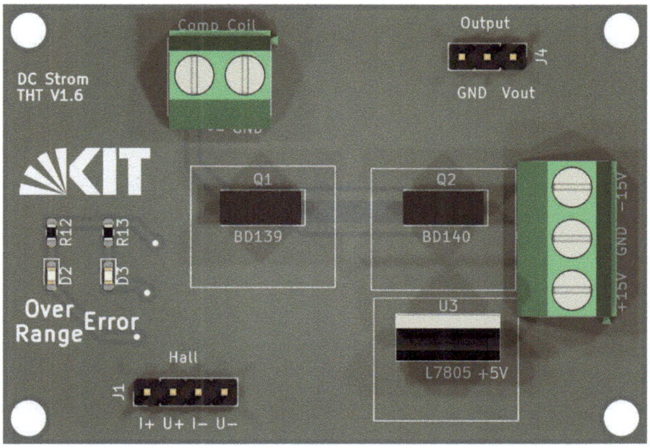

(a) front

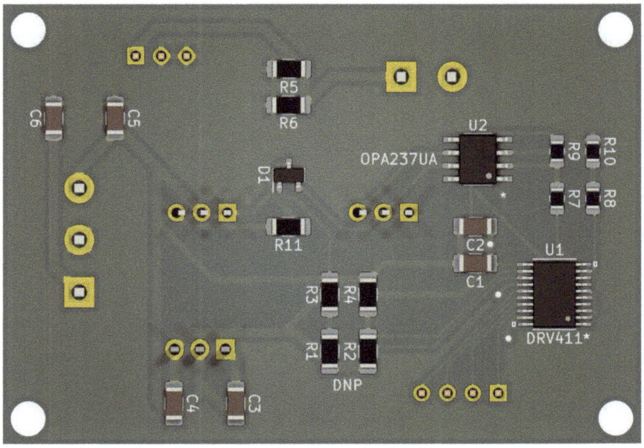

(b) back

Figure A.1: Front and back of the PCB for closed-loop hall effect current sensor.

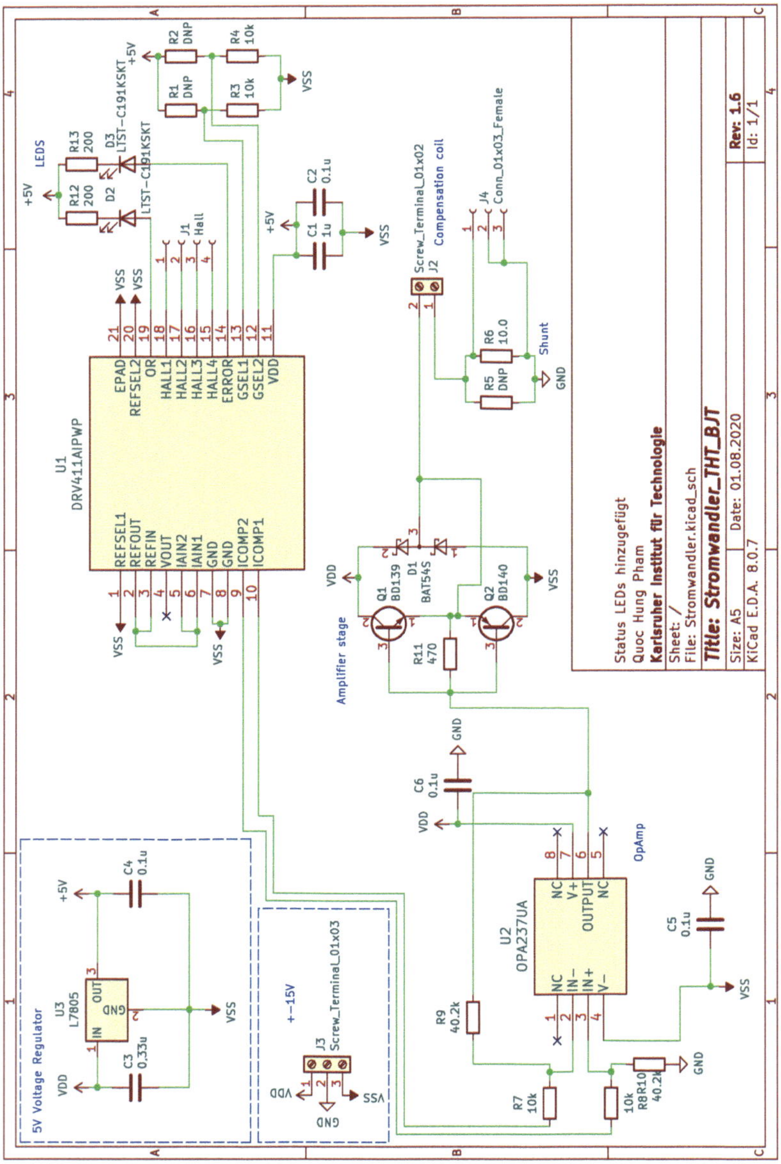

Figure A.2: Schematic of the current transducer electronics

Table A.1: Bill of materials of the current transducer electronics

Reference	Description	Qty
C1	1 µF, 1206, Ceramic, X7R, 50 V, 10 %, Capacitor	1
C2, C4, C5, C6	0.1 µF, 1206, Ceramic, X7R, 50 V, 10 %, Capacitor	4
C3	0.33 µF, 1206, Ceramic, X7R, 50 V, 10 %, Capacitor	1
R3, R4	10 kΩ, 1206, 0.25 W, 1 %, Resistor	2
R6	10 Ω, 1206, 0.25 W, 0.1 %, Resistor	1
R7, R8	10 kΩ, 0805, 0.1 W, 0.1 %, Resistor	2
R9, R10	40.2 kΩ, 0805, 0.1 W, 0.1 %, Resistor	2
R11	470 Ω, 1206, 0.25 W, 1 %, Resistor	1
R12, R13	200 Ω, 0603, 0.1 W, 1 %, Resistor	2
D1	BAT54S, Schottky Diode	1
D2, D3	LTST-C191KSKT, LED green	2
U1	DRV411	1
U2	OPA237UA	1
U3	L7805, Voltage regulator	1
Q1	BD139, NPN	1
Q2	BD140, PNP	1

B Material Properties

B.1 Resistivities

Resistivities of the materials that appear in HTS tapes.

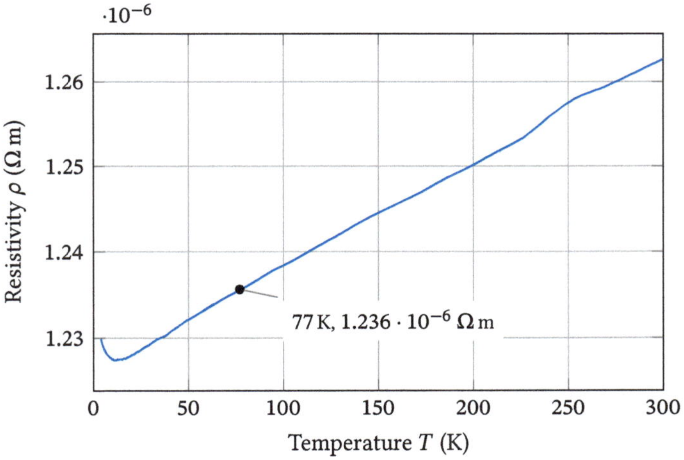

Figure B.1: Resistivity of Hastelloy in the temperature range from 4 K to 300 K based on [Lu08].

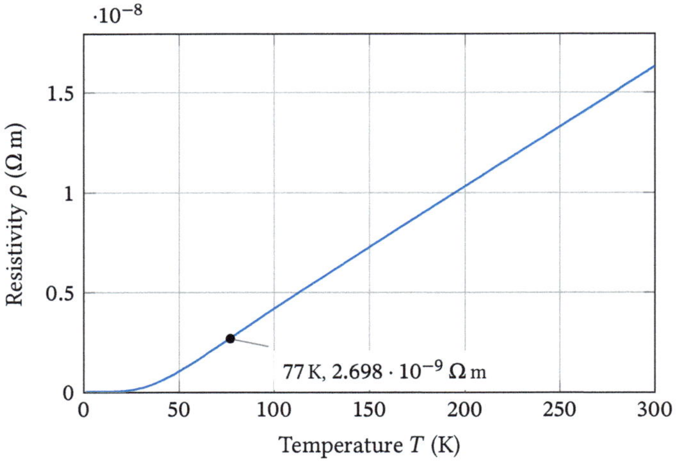

Figure B.2: Recommended resistivity of silver in the temperature range from 4 K to 300 K based on [Mat79].

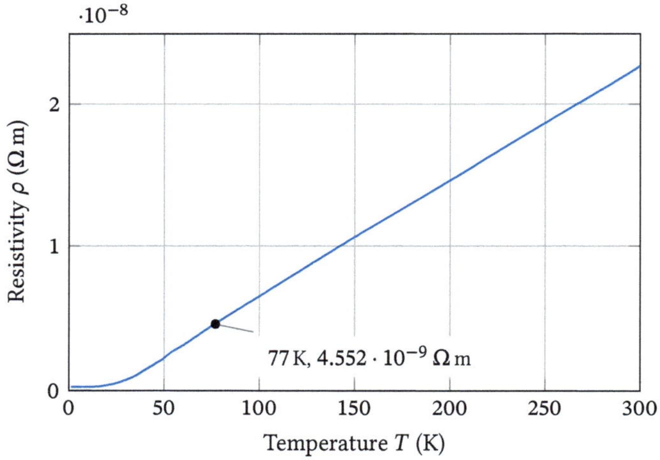

Figure B.3: Recommended resistivity of gold in the temperature range from 4 K to 300 K based on [Mat79].

B.2 HTS Tape Geometry

SEM images of the superconductor used in this work. Due to the high magnification, the layer thicknesses of the individual layers can be determined.

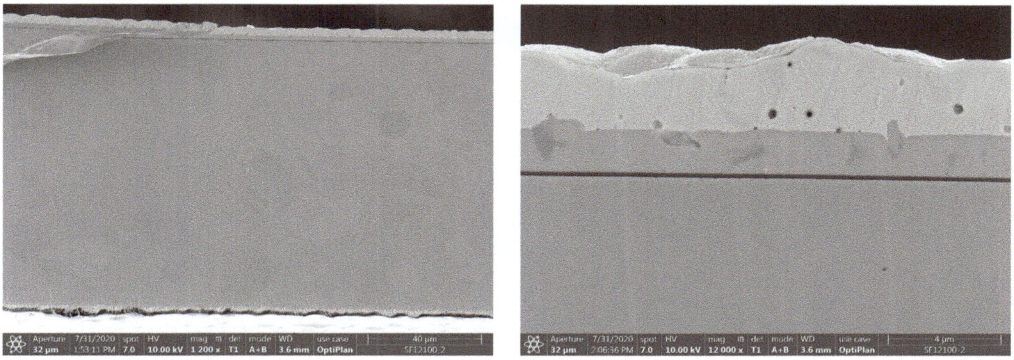

(a) (b)

Figure B.4: SEM of the middle section of a superconducting tape Superpower SF12100 with a magnification of 1200 (a) and 12000 (b).

B.3 Liquid Nitrogen

Boiling curve to determine the excess temperature ΔT at a given heat flux density $\dot{q}$ under pool boiling conditions in LN_2.

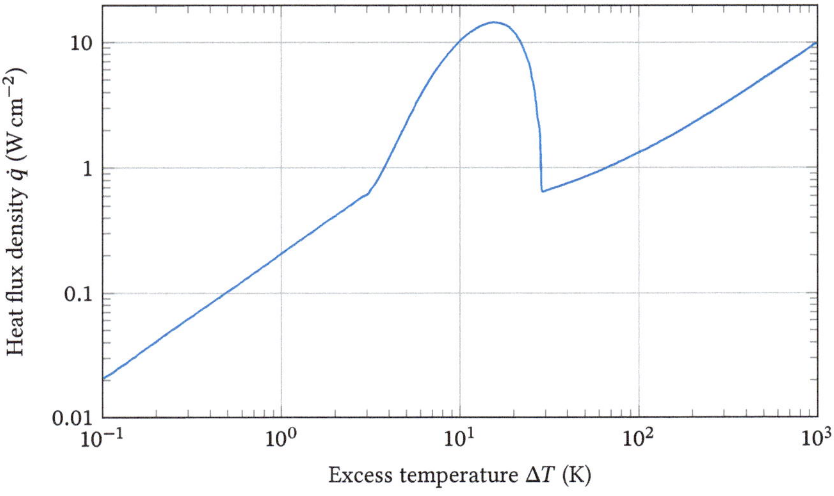

Figure B.5: Boiling curve and heat flux density of LN_2 at ambient pressure based on [Mer64].

C Mathematical Functions

C.1 Exponential Decay and Increase

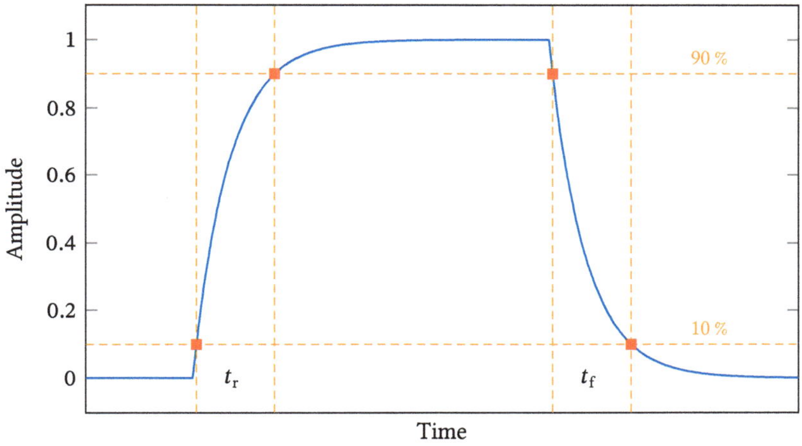

Figure C.1: Rise and fall time of an exponential increase and decay.

Assuming an exponential decay or rise

$$V(t) = V_{max} e^{-\frac{t}{\tau}} \quad (C.1)$$

$$V(t) = V_{max} \left(1 - e^{-\frac{t}{\tau}}\right) \quad (C.2)$$

the 10 % to 90 % rise or fall time $t_{r/f}$ is proportional to the exponential time constant τ

$$t_{r/f} = (\ln 0.9 - \ln 0.1) \cdot \tau \approx 2.1972 \cdot \tau \quad (C.3)$$

$$\tau = \frac{t_{r/f}}{\ln 0.9 - \ln 0.1} \approx \frac{t_{r/f}}{2.1972} \quad (C.4)$$

Karlsruher Schriftenreihe zur Supraleitung

Karlsruher Institut für Technologie (KIT) | ISSN 1869-1765

Band 001 **Christian Schacherer**
Theoretische und experimentelle Untersuchungen zur Entwicklung supraleitender resistiver Strombegrenzer.
ISBN 978-3-86644-412-6

Band 002 **Alexander Winkler**
Transient behaviour of ITER poloidal field coils.
ISBN 978-3-86644-595-6

Band 003 **André Berger**
Entwicklung supraleitender, strombegrenzender Transformatoren.
ISBN 978-3-86644-637-3

Band 004 **Christoph Kaiser**
High quality Nb/Al-AlOx/Nb Josephson junctions. Technological development and macroscopic quantum experiments.
ISBN 978-3-86644-651-9

Band 005 **Gerd Hammer**
Untersuchung der Eigenschaften von planaren Mikrowellenresonatoren für Kinetic-Inductance Detektoren bei 4,2 K.
ISBN 978-3-86644-715-8

Band 006 **Olaf Mäder**
Simulationen und Experimente zum Stabilitätsverhalten von HTSL-Bandleitern.
ISBN 978-3-86644-868-1

Band 007 **Christian Barth**
High Temperature Superconductor Cable Concepts for Fusion Magnets.
ISBN 978-3-7315-0065-0

Die Bände sind unter www.ksp.kit.edu als PDF frei verfügbar oder als Druckausgabe bestellbar.

Band 008 **Axel Stockhausen**
Optimization of Hot-Electron Bolometers for THz Radiation.
ISBN 978-3-7315-0066-7

Band 009 **Petra Thoma**
Ultra-fast $YBa_2Cu_3O_{7-x}$ direct detectors for the THz frequency range.
ISBN 978-3-7315-0070-4

Band 010 **Dagmar Henrich**
Influence of Material and Geometry on the Performance of Superconducting Nanowire Single-Photon Detectors.
ISBN 978-3-7315-0092-6

Band 011 **Alexander Scheuring**
Ultrabreitbandige Strahlungseinkopplung in THz-Detektoren.
ISBN 978-3-7315-0102-2

Band 012 **Markus Rösch**
Development of lumped element kinetic inductance detectors for mm-wave astronomy at the IRAM 30 m telescope.
ISBN 978-3-7315-0110-7

Band 013 **Johannes Maximilian Meckbach**
Superconducting Multilayer Technology for Josephson Devices.
ISBN 978-3-7315-0122-0

Band 014 **Enrico Rizzo**
Simulations for the optimization of High Temperature Superconductor current leads for nuclear fusion applications.
ISBN 978-3-7315-0132-9

Band 015 **Philipp Krüger**
Optimisation of hysteretic losses in high-temperature superconducting wires.
ISBN 978-3-7315-0185-5

Band 016 **Matthias Hofherr**
Real-time imaging systems for superconducting nanowire single-photon detector arrays.
ISBN 978-3-7315-0229-6

Band 017 **Oliver Näckel**
Development of an Air Coil Superconducting Fault Current Limiter.
ISBN 978-3-7315-0526-6

Band 018 **Christoph M. Bayer**
Characterization of High Temperature Superconductor Cables for Magnet Toroidal Field Coils of the DEMO Fusion Power Plant.
ISBN 978-3-7315-0605-8

Band 019 **Shengnan Zou**
Magnetization of High Temperature Superconducting Trapped-Field Magnets.
ISBN 978-3-7315-0715-4

Band 020 **Ilya Charaev**
Improving the Spectral Bandwidth of Superconducting Nanowire Single-Photon Detectors (SNSPDs).
ISBN 978-3-7315-0745-1

Band 021 **Juliane Raasch**
Electrical-field sensitive $YBa_2Cu_3O_{7-x}$ detectors for real-time monitoring of picosecond THz pulses.
ISBN 978-3-7315-0786-4

Band 022 **Yingzhen Liu**
Design of a superconducting DC wind generator.
ISBN 978-3-7315-0796-3

Band 023 **Sebastian Hellmann**
Research and Technology Development on Superconducting Current Limiting Transformers.
ISBN 978-3-7315-0804-5

Band 024 **Simon J. Otten**
Characterisation of REBCO Roebel cables.
ISBN 978-3-7315-0904-2

Band 025 **Julia Brandel**
Supraleitende Einzelphotonenzähler: Optimierung der Zeitauflösung und Anwendungsbeispiele aus der Spektroskopie.
ISBN 978-3-7315-0917-2

Band 026 **Dustin Kottonau, Eugen Shabagin, Wescley T. B. de Sousa, Jörn Geisbüsch, Mathias Noe, Hanno Stagge, Simon Fechner, Hannes Woiton, Thomas Küsters**
Bewertung des Einsatzes supraleitender 380-kV-Kabel.
ISBN 978-3-7315-0927-1

Band 027 **Steffen Dörner**
Multifrequenzausleseverfahren von supraleitenden Einzelphotonen-Detektoren.
ISBN 978-3-7315-0961-5

Band 028 **Michael Merker**
Superconducting integrated THz receiver.
ISBN 978-3-7315-0970-7

Band 029 **Wolfgang-Gustav Ekkehart Schmidt**
Superconducting Nanowire Single-Photon Detectors for Quantum Photonic Integrated Circuits on GaAs.
ISBN 978-3-7315-0980-6

Band 030 **Dustin Kottonau, Eugen Shabagin, Wescley de Sousa, Jörn Geisbüsch, Mathias Noe, Hanno Stagge, Simon Fechner, Hannes Woiton, Thomas Küsters**
Evaluation of the Use of Superconducting 380 kV Cable.
ISBN 978-3-7315-1026-0

Band 031 **Alan Preuß**
Development of high-temperature superconductor cables for high direct current applications.
ISBN 978-3-7315-1041-3

Band 032 **Roland Gyuráki**
Fluorescent thermal imaging method for investigating transient effects in high-temperature superconductor tapes and coils.
ISBN 978-3-7315-1064-2

Band 033 **Aurélien Godfrin**
AC Losses in High-Temperature Superconductor Tapes and Cables for Power Applications.
ISBN 978-3-7315-1096-3

Band 034 **Alexander Schmid**
Integrierte planare Antennen für supraleitende Detektorsysteme zur THz-Strahldiagnostik.
ISBN 978-3-7315-1145-8

Band 035 **Alexander Buchholz**
Prospective Life Cycle Assessment of High-Temperature Superconductors for Future Grid Applications.
ISBN 978-3-7315-1194-6

Band 036 **Emanuel Marius Knehr**
Technology and readout for scaling up superconducting nanowire single-photon detectors.
ISBN 978-3-7315-1256-1

Band 037 **Mathias Noe, Chistof Humpert, Wescley de Sousa, Stefan Fink, Friedhelm Herzog, Stefan Huwer, Marvin Nießen, Sebastian Pfaller, Stephan Pöhler, Tobias Rodler, André Schmid, Tabea Arndt**
380 kV Superconducting Fault Current Limiter Feasibility Study.
ISBN 978-3-7315-1313-1

Band 038 **Carsten Räch**
Untersuchung von Wechselstromverlusten verschiedener hochtemperatursupraleitender Mehrleiterkonzepte für dreiphasige Hochstromanwendungen.
ISBN 978-3-7315-1372-8

Band 039 **Ali Khonya**
System-Level Modeling of the Powertrain for an Electric Aircraft with Superconducting Components.
ISBN 978-3-7315-1435-0

Band 040 **Quoc Hung Pham**
Development of a Superconducting Inverter Based on Magnetic Field Triggered Dynamic Resistance.
ISBN 978-3-7315-1452-7